Physical Oceanography of the Dying Aral Sea

Peter Zavialov

Physical Oceanography of the Dying Aral Sea

 Springer

Published in association with
Praxis Publishing
Chichester, UK

Dr Peter Zavialov
Shirshov Institute of Oceanology
Moscow
Russia

SPRINGER–PRAXIS BOOKS IN GEOPHYSICAL SCIENCES
SUBJECT *ADVISORY EDITOR*: Dr Philippe Blondel, C.Geol., F.G.S., Ph.D., M.Sc., Senior Scientist, Department of Physics, University of Bath, Bath, UK

ISBN 3-540-22891-8 Springer-Verlag Berlin Heidelberg New York

Springer is part of Springer-Science + Business Media (springeronline.com)

Bibliographic information published by Die Deutsche Bibliothek

Die Deutsche Bibliothek lists this publication in the Deutsche Nationalbibliografie; detailed bibliographic data are available from the Internet at http://dnb.ddb.de

Library of Congress Control Number: 2004117081

Cover design: Jim Wilkie
Project Management: Originator Publishing Services, Gt Yarmouth, Norfolk, UK

Printed on acid-free paper

Contents

Preface

The objectives of this book are twofold. On the one hand, it is aimed at describing the present physical state of the Aral Sea (Chapter 3) as well as identifying likely future scenarios (Chapter 4) based on the data recently collected from the lake. On the other hand, it is intended to provide a concise summary of the existing literature, both recent and "old" (i.e., dedicated to the hydrology of the Aral Sea prior to the ongoing shrinking or at the early stages of the desiccation) (Chapter 2). An overview of the historical background is given in Chapter 1, and some implications of the Aral crisis viewed as part of the global perspective are discussed in Chapter 5. Accordingly, the book combines the information excerpted from the literature (a large part of which was published in Russian in the former USSR and, therefore, has not been readily available to the international reader), with new results. The latter include, in particular:

- a description of the 3D structure of thermohaline fields in Large Aral and some physical properties of Aral's waters in 2002–2004 (Sections 3.4 and 3.8);
- temperature–salinity (TS) analyses of water types in the present Aral and quantitative estimates of salt and mass exchanges between the eastern and western basins of Large Aral (Section 3.5);
- new information about the circulation of the present Aral derived from direct current measurements, modeling, and remote sensing (Section 3.6);
- data quantifying the anoxic conditions and H_2S content in the Aral Sea (Section 3.7);
- a numerical assessment of plausible future scenarios, depending on the river discharges and evaporation rates (Chapter 4).

A part of these recent results has not been previously published.

The present Aral Sea is a rapidly changing and underexplored environment, and the author fully recognizes that the proposed description is incomplete in many respects. Some concepts might need to be revisited as the research progresses. The goal was to provide a topical account to date.

As made explicit by the book title, we deliberately restrict the content to the descriptive physical oceanography of the Aral Sea as a large inland water body, and immediately adjacent topics. The other broad and, undoubtedly, important aspects of the Aral Sea science, such as those related to biological and ecological problems, socio-economic and health consequences, land hydrology and water management, etc., are either only briefly mentioned or not mentioned at all. These themes have been addressed elsewhere. The book is not intended as an exhaustive bibliographic review, although a number of the most important references are given.

An important constituent of the book is the information collected during 4 field surveys in 2002–2004. These expeditions were organized by the Russian Academy of Sciences through the Shirshov Institute of Oceanology and its Laboratory of Experimental Physical Oceanography. I would like to thank the Laboratory and the Institute and personally thank S. Lappo, A. Zatsepin, and M. Flint for support. The research was also funded by grants from the Russian Foundation for Basic Research, the Russian Ministry of Science and Technology, NATO, the National Geographic Society (USA), and by the Academy of Sciences of Uzbekistan. I also thank A. Ambrosimov, F. Amirov, S. Dikarev, J. Friedrich, A. Ginzburg, D. Ishniyazov, V. Khan, A. Kostianoy, T. Kudyshkin, A. Kurbaniyazov, S. Kurbaniyazov, B. Levin, A. Ni, D. Nourgaliev, H. Oberhänsli, M. Petrov, D. Rodin, S. Stanichniy, F. Sapozhnikov, N. Sheremet, O. Stroganov, A. Subbotin, V. Zhurbas, and other colleagues and friends in Russia, Uzbekistan, Kazakhstan, Germany, Ukraine, Belgium, and elsewhere, whose invaluable collaboration in one or another form has made this book possible.

About the author

Peter O. Zavialov graduated from the Department of Physics, Moscow State University in 1989. He received his Candidate of Sciences degree (considered the Russian equivalent to a PhD) in physics and mathematics, specializing in physical oceanography, from the State Oceanographic Institute, Moscow, in 1992, and his Doctor of Sciences degree in geography (oceanography) from Shirshov Institute of Oceanology, Moscow, in 2000. His current position is Leading Research Scientist at Shirshov. Peter has published about 60 research papers focused on sea–air–land interactions, remote sensing techniques, shelf circulations in the South Atlantic, anthropogenic impacts and global change, and the present state of the Aral Sea. He is also the author of several popular articles. Peter served as a Principal Investigator in a number of field surveys in Aral in the last few years.

Figures

Tables

Abbreviations

3D	3-Dimensional
AN SSSR	Academy of Sciences (USSR)
a.o.l.	above ocean level
ARW	Advanced Research Workshop
ASI	Advanced Science Institute
AVHRR	Advanced Very High Resolution Radiometer
BP	Before Present
CTD	Conductivity–Temperature–Depth
GOIN	State Oceanographic Institute (USSR, Russia)
ICWC	International Coordination Water Commission (Central Asia)
ILEC	International Lake Environment Committee
IOA	International Ocean Thermal Energy Conversion Association
KazNIGMI	Kazakhstan Institute of Hydrometeorology
MCSST	Multi-Channel Sea Surface Temperature
MCC	Maximum Cross-Correlation technique
NATO	North Atlantic Treaty Organization
NCAR/NCEP	National Center for Atmospheric Research/National Center for Environmental Prediction (USA)
NOAA	National Oceanic and Atmospheric Administration (USA)

POM   Princeton Ocean Model

RegCM   Regional Climate Model

SANIGMI   Central Asian Hydrometeorological Institute (Uzbekistan)
SANII   Central Asian Scientific Research Institute (USSR)
SST   Sea Surface Temperature

TS   Temperature–Salinity

UNESCO   United Nations Educational, Scientific and Cultural
                 Organization
USSR   Union of Soviet Socialist Republics

VNIIRO   All-Union Institute of Fisheries and Oceanography (USSR)

Introduction

I.1 THE ARAL SEA

At the time of writing (2004), the Aral Sea, formerly one of the largest lakes on Earth, an oasis surrounded by Central Asian deserts, has lost 75% of its surface area and about 90% of its water. The newly dry bottom occupies an area exceeding the territory of Belgium in a remote region of Kazakhstan and Uzbekistan, now independent republics of the former Soviet Union. Aralsk and Muynak, busy and wealthy harbor cities of the 1950s, are now located tens of kilometers away from the present shoreline. A fleet of fishing and cargo ships which once cruised the brackish Aral Sea waters, now rests on the former bottom, quickly disappearing under rust, salt, and sands. Only a few decades ago, biological communities of the Aral Sea and the adjacent deltaic areas included hundreds of species, some of which were endemic. Fishery yields were as large as up to 50,000 tons per year, making up a considerable part of the fish catches of the USSR. For example, the Aral Sea accounted for up to 13% of sturgeon catches. The cargo freight turnover was over 200,000 tons per year. By the 1980s, commercial fishery and navigation had ceased completely, as efforts to keep the ports open became too difficult and expensive (e.g., Micklin, 2004).

It is often thought that the cause of the Aral Sea shrinking was purely anthropogenic. The shallowing was, undoubtedly, triggered by unsustainable diversions of water resources for irrigating the cotton and rice plantations. But many specialists believe that the desiccation is also, at least partly, due to larger scale processes of natural climate variability. However, the contemporary Aral desiccation is considered the world's worst aquatic ecology crisis in recent history. Negative effects of the Sea's retreat on the economy, ecology, and quality of human life in the region were manifold and dramatic, as described in the broad literature.

The title of this book refers to Aral as a "dying sea", following a cliché recurrently appearing in recent publications. But is Aral really dying? There is a wealth of geological and archaeological evidence indicating that similar or even stronger regressions of the Sea have occurred in the distant past, always followed by subsequent recoveries. Today, the Sea has not vanished, it has just drastically changed and fled farther into the deserts, still rather deep and seemingly boundless, beautiful and blue (Figure I.1, see colour section), and largely unexplored in its new capacity. In this book, we focus on changing physical properties of the Aral Sea as a large and special water body from the oceanographic standpoint. We deliberately apply this term to an object whose spatial scales are definitely not oceanic, not even a "real" sea—it has been long known that the Aral Sea uniquely combined lacustrine and marine properties and admitted many oceanographic approaches to research.

The importance of studying the present critical physical state of the Aral Sea is dual. First, the problem is of obvious intrinsic significance, given that any reasonable forecast of further changes of the Aral Sea must rely on an accurate description and correct understanding of the present conditions. Needless to say, a sound prediction is of great applied importance, as it may help to elaborate measures aimed at easing the consequences of desiccation. Second, the present Aral Sea, with its very special oceanographic environment, is a unique natural "laboratory" which can be used to investigate physical and chemical processes taking place in other lakes, inland seas, and even oceanic regions, but manifested in the Aral Sea in their extreme form. Relevant general issues include, but are not limited to, mixing and turbulence in strongly stratified geophysical fluids, salinity-controlled circulations in hyperhaline water bodies, oxygen depletion and hydrogen sulphide production in anoxic zones, etc. The processes presently underway in the Aral Sea may be quite instructive with respect to a number of other lakes which are experiencing similar desiccation or are otherwise endangered (e.g., Lake Chad or the Dead Sea).

I.2 WHY WRITE THIS BOOK?

Until the early 1990s, the physical regime of the Aral Sea was subject to extensive monitoring by means of regular research cruises conducted several times a year, continuous routine observations at up to 11 nearby stationary hydrometeorological stations, aircraft surveys made 2–12 times a year, and other monitoring methods. At the time, the Aral Sea was one of the most well-explored seas washing the former USSR. The results were published in hundreds of research papers and a number of books. Many Russian-speaking researchers interested in the Aral Sea keep the books by L.K. Blinov (1956), A.N. Kosarev (1975), I.V. Rubanov et al. (1987), and other distinguished authors on their desktops. At the end of this "golden era" of the Aral research, a collective volume edited by V.N. Bortnik and S.P. Chistyaeva (1990) was published. Many important previous results were reviewed and concisely summarized in this fundamental work.

Unfortunately, most of these publications have never been translated into English. Almost all relevant articles were published in Russian in Soviet periodicals or, in some cases, as internal reports of research units, and therefore are not readily available to the international scientific community. For example, of the 273 biblio-graphic citations given in the book edited by Bortnik and Chistyaeva (1990), all were in Russian. Extremely few, if any, Aral related papers were published by foreign authors by the beginning of the 1990s, which is explicable, given that the Aral Sea was an internal sea of the USSR. Moreover, foreign researchers were not very welcome in this particular region because of the presence of secret military facilities, including a bacteriological weapon testing site at Vozrozhdeniya Island. The situation has changed following the end of the Cold War, but a certain imbalance towards Russian language literature is still evident now. According to the most comprehensive Aral Sea bibliography volume by Nihoul et al. (2002), out of a total of 1,540 related bibliographic items published before 2000, only about 300 were in English, with only a few in German and French. A few well-known books on Aral, such as *Létolle and Mainguet* (1993), for example, and several volumes of collected articles have been subsequently published abroad in the 1990s, but most of those were not focused on physical oceanographic issues, mainly addressing other important facets of the Aral problem.

In the early 1990s, the amount of scientific information obtained from the Aral Sea had decreased significantly. This was partly because of the well-known political and economical challenges facing the region after the decay of the USSR, accom-panied by a temporary general depression of scientific research in all former Soviet republics. In addition, by that time, the shoreline had retreated too far away from the infrastructure long used for ship operations (formerly situated mainly in the cities of Aralsk and Muynak), roads, populated settlements, and potable water sources. The Sea's water body had been encircled by thousands of square kilometers of the former sea bottom which, at many locations, was virtually impassable even by off-road vehicles. Because of all these factors, any field research in the Sea became technically and logistically much more difficult, and also increasingly expensive at the time when local financial resources were particularly short. As far as we know, the last reasonably large scale shipborne survey in Large Aral was undertaken in 1992. Despite continued selfless efforts of several research groups and individuals, the hydrophysical and hydrochemical data for the subsequent decade were very sparse. In our opinion, the decade of the 1990s was a "phase transition" time when the cumulative changes of the preceding desiccation period have resulted in a qualitatively new physical state of the lake. It is unfortunate that this interesting period in the recent evolution of the Aral Sea is perhaps the least explored, as far as the physical oceanographic processes are concerned. We know of a number of small-scale expeditions to the Sea conducted during this period, but most of them had no means to provide the necessary 3D hydrographic data. However, determinations of surface salinity and some other basic measurements, at least at a few near-shore locations, have been repeatedly undertaken, and it is thanks to these works that we have an idea of the salinization progress in the 1990s. Of course, important informa-tion has also been gathered by means of satellite imagery in different spectral bands, but the *in situ* data were needed. The situation seems to have started to change for the better in the early 2000s, with an increasing number of national and international field activities. These studies have yielded new hydrographic data at a reasonable accuracy and coverage, but it should be kept in mind that the present Aral Sea is still underexplored. From the oceanographic standpoint, today's Aral Sea is a very special object which has little in common with the "old" Sea. By now, the Aral Sea problem is at its peak of attention from the worldwide scientific community, and a considerable amount of new information has been obtained through direct measurements, remote sensing, and modeling. These new findings are scattered among numerous publications.

Hence, the motivation for writing this book is as follows. First, we believe that it is time to make an attempt to summarize and generalize recent results to provide a topical look at the physical state of the present Aral Sea, and also to sketch its possible future prospects. Secondly, the book is intended to provide at least a brief summary of books and articles describing the physical oceanographic state of the "old", pre-desiccation Aral Sea previously published in Russian and thus partly fill a gap existing in the English language literature, making this information accessible for the international reader.

1

A brief historical overview

We begin with an overview of the paleovariability of the Aral Sea and its regressions and transgressions in the distant past, and then discuss the historical background focusing on human activities in the Aral Sea region and exploration of the lake. Either subject has been discussed in detail in the broad literature, and we restrict this chapter to only an introductory account.

The geographic location of the Aral Sea is shown in Figure 1.1. The depression presently occupied by the Sea, together with the nearby Sarakamysh and Khorezm depressions located south of the Aral Sea, was formed by tectonic activity on the Turan Plate in the late Neogene (e.g., Pinkhasov, 2000) and, subsequently, metamorphized by wind erosion and river alluvia. In the late Pleistocene, the Aral depression was a dry (maybe except for some areas occupied by salty marshes) plain with indented relief (Ashirbekov and Zonn, 2003). Later, the waters of the Syr-Darya River partly filled the basin, forming a lake of moderate size whose surface is believed to have been about 31 m above the ocean level (a.o.l.). Some investigators of Aral's bottom sediments have described a terrace at this elevation, presumably associated with the lake's surface standing during the period (e.g., Vaynbergs and Stelle, 1980). At the time, the regional climate was cold and dry. According to some paleoreconstructions, the annual fluvial discharge into the lake was between 8 and 10 km^3 on average (Mamedov and Trofimov, 1986) (i.e., 15–20% of that characteristic for the Aral Sea in the mid-20th century).

It is believed that at these early stages of the Aral Sea's history, the precursor of the Amu-Darya River ran not into the Aral Sea but into the Caspian Sea, merging with it south of the present Kara-Bogaz-Gol Bay. Sometime in the Pleistocene or early Holocene, Amu-Darya drastically changed its course and turned northward to the Aral depression, leaving on its former course toward the Caspian Sea extensive alluvia and ancient dry beds in a valley presently known as Uzboy. Neither the time nor the dynamical mechanism responsible for such a change in the course of the Amu-Darya River are known exactly. The available dating estimates for the Sea

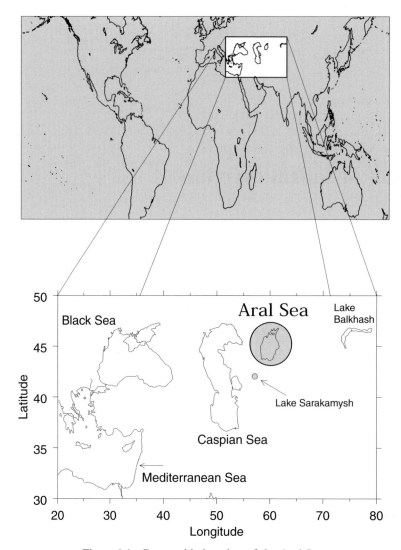

Figure 1.1. Geographic location of the Aral Sea.

filling vary in a broad range from 9,000–10,000 years before present (BP) (Kvasov, 1991; Aladin and Plotnikov, 1995b), 15,000–17,000 years BP (Nurtaev, 2004), and even up to 24,000 years BP (e.g., Pshenin et al., 1984; Rubanov et al., 1987). Tectonic activity in the region is believed to have been the principal dynamical origin of Amu-Darya's turn. We note apropos that, in terms of geodynamics, the present Aral region can be divided into two distinct domains. According to the recent instrumental measurements reported by Nurtaev (2004), the domain south of 45°N presently exhibits strong tectonic uplift at rates of 5–12 mm/year, while the northern domain shows little or no vertical displacement.

As shown by paleoclimate reconstructions, the regional climate has changed to warmer and wetter conditions in the early to mid-Holocene, and the Aral Sea, already fed by Amu-Darya and Syr-Darya together, has been receiving river runoff at the rate of about $150 \, km^3$/year on average (Mamedov and Trofimov, 1986) (i.e., almost 300% of the runoff typical to the 20th century prior to the latest desiccation). The corresponding transgression of the Sea has led to the flooding of the Sarakamysh and Khorezm depressions. According to a widespread conception, the lake surface has reached an elevation of 57–60 m a.o.l., and then Aral water has spilled into Uzboy through the southern extremity of Sarakamysh. By most plausible estimates, this happened sometime between 3,000 and 8,000 years BP. As a result, Amu-Darya regained the connection to the Caspian Sea while still feeding the Aral Sea. The annual discharge into the Caspian Sea through Uzboy during this period has been estimated as $60–80 \, km^3$ (Mamedov and Trofimov, 1986). Terraces at 58–60 m above the ocean level seen at the south-eastern shore of the lake are interpreted as visible evidence of this ancient Aral standing (e.g., Vaynbergs and Stelle, 1980). A number of authors also described terraces at 72–73 m and even 75–80 m a.o.l. (e.g., Gorodetskaya, 1978) and discussed the possibility of the lake standing at these high levels sometime before 5,000 years BP (e.g., Fedorov, 1980; Aladin and Plotnikov, 1995b; Boomer et al., 2000). Such a level would imply that the lake's area was greater then $150,000 \, km^2$ (i.e., more than twice the area of the pre-desiccation Aral in the 20th century). However, there is no general agreement in the literature about the age of the terraces, nor is it clear to what extent their present elevations are representative of the corresponding historical lake levels. As discussed by Rubanov et al. (1987), the position of the terraces may have been significantly altered by subsequent tectonic movements. Moreover, recent archeological findings reportedly rule out lake standings above 60 m (Baypakov et al., 2004)

One of the regressions of the Aral Sea occurred 3,200–3,800 years BP (e.g., Nurtaev, 2004), after the regional climate had changed once again towards drier conditions. Consequently, the outflow into the Caspian Sea ceased and the lake level dropped as low as 35 m a.o.l. (Rubanov et al., 1987). The regression was followed by a recovery, and the Aral Sea level is believed to have varied between 45 m and 55 m a.o.l. until about 1,500–1,900 years BP, when a new deep regression took place (Nurtaev, 2004). According to some data, this regression was the strongest on record. The lake level dropped to 27–28 m a.o.l., evident from the layers of gypsum and mirabilite sedimented at the time. As shown by Rubanov and Timokhina (1982), the salinity must have exceeded 150–160 ppt during such regressions. Following a period of recovery, the next strong regression occurred 450–800 years BP. At the time, the minimum Aral Sea level was about 31–35 m a.o.l., and a gypsum layer was sedimented. This most recent of the regressions preceding the present desiccation seems to be well documented not only by geological but also by historiographic and archaeological evidence.

Some authors have argued that a major part of the Amu-Darya runoff drained into Sarakamysh rather than Aral during the last two deep regressions, and Lake Sarakamysh, whose volume greatly increased, was again connected with the Caspian Sea through Uzboy during these events. According to Aladin and Plotnikov (1995b), the medieval regression ended in late 16th century, after the main Amu-Darya delta had moved from Lake Sarakamysh to the Aral Sea, thus resulting in rapid growth of Aral and a shrinking of Sarakamysh. We note that the Aral Sea shallowing in the 20th century has been accompanied by growth and deepening of Lake Sarakamysh whose present depth is about 40 m (Ashirbekov and Zonn, 2003). Some of the water diverted from Amu-Darya and hence withdrawn from Aral's basin is eventually dumped into Sarakamysh, at an estimated rate of 4–5 km per year. It can be said, therefore, that the volumes of the two lakes have often varied in an anti-correlated pattern.

Thus, the Aral Sea has undergone a series of major desiccation and flooding episodes in the more or less distant past. Although the paleovariability of the Aral Sea has been forced by natural climate changes, some researchers hypothesized that the level changes over the historical period were at least partly anthropogenic (e.g., Kvasov and Mamedov, 1991), considering that the Aral Sea region (the ancient Khorezm) has been an area of extensive irrigation, with the total area of irrigated lands totaling up to about 15,000 km^2 in the lower reaches of Amu-Darya and 25,000 km^2 in those of Syr-Darya (Ashirbekov and Zonn, 2003). There is solid evidence indicating that the people of Khorezm built hydrotechnical installations such as dams and canals and were able to regulate the river discharges to a certain extent (e.g., to distribute the Amu-Darya runoff between Aral and Sarakamysh). Some authors have argued, however, that the irrigated agriculture in itself has played only a minor role in Aral's water budget, and the ancient anthropogenic impacts were mainly associated with frequent wars or social disorders which sometimes resulted in destruction of the dams (e.g., Ashirbekov and Zonn, 2003).

According to V.V. Bartold, one of the most authoritative scholars of Central Asia's history, virtually the first written allusions of the Aral Sea are found in Chinese sources in the 2nd century BC, where the Sea is mentioned in rather loose terms and referred to as the "Northern Sea" or "Western Sea" (Barthold, 1902). Some earlier Roman sources also mention "Oxian marshes" (*Palus Oxiana*) in the low reaches of Amu-Darya commonly called Oxus at the time. More elaborate information can be found in the writings of arabic geographers of the 10th, 11th, and 12th centuries. The Sea is described as Amu-Darya's terminal salty lake, whose size and shoreline contour are rather close to those known in the mid-20th century. It was also reported in these sources that the Sea had no connection with Sarakamysh. A traveler who wanted to go from Khorezm to the "land of Pechenegs" (apparently the lower Volga and Urals regions of present Russia) was advised to climb the "Khorezmian mountains" (i.e., the Ustyurt Plateau cliff bordering the western shore of the lake), and then go north through a "waterless desert", leaving the Aral Sea ("Khorezmian Lake" or "Jend Lake") on his right. We note that these directions describe quite well the shortest way from Uzbekistan to Russia, frequently used today by drivers of all-terrain trucks and vehicles.

There were little or no written mentions of the Aral Sea between the early 13th and late 16th centuries, the medieval regression period. Moreover, some sources of 15th century origin have claimed that the "Khorezmian Lake" known from "ancient books" no longer existed at the time. Syr-Darya was thought to either merge into

Amu-Darya somewhere, or even disappear in the sandy desert, and Amu-Darya was believed to be a tributary of the Caspian Sea. There is evidence showing that Uzboy was, indeed, filled with water and even used for navigation in the 14th century (Barthold, 1902). In the 17th century, Abulgazi (1603–1664), khan of Khiva and historian characterized as an "enlightened and educated man" by his contemporaries, describes the Sea as a terminal lake of Syr-Darya and called it "Sea of Syr" (*Syr Tengizi*). According to Abulgazi, Amu-Darya had regained its connection to the Aral Sea only in 1572–1573.

The word "Aral" meaning "island" in Turkic was initially used by Abulgazi to identify the deltaic area of Amu-Darya. During a few decades in the 17th century, the region was a sovereign state independent of the Khiva khanate, with the capital in Kungrad.

The name "Aral Sea" first appeared in Russian texts in 1697 (in some earlier Russian sources, the Sea appears as *Sineye More* (i.e., "Blue Sea")) and was adopted in European maps in the 1720s. The expansion of the Russian empire into the region and the eventual establishment of a Russian protectorate over the Kokand and Khiva khanates and Bukhara emirate (1873) opened a new page in the recent history of Aral. The first permanent Russian fortified settlement on the Aral Sea, Fort Raim and its shipyard, was erected in 1847 on Syr-Darya, not far from its mouth. Already in 1869, the Russian authorities started prospecting and works aimed at constructing new irrigation facilities. A detailed account and chronology of the irrigation-related activities are presented by, for example, Ashirbekov and Zonn (2003). In particular, a canal from Syr-Darya to Golodnaya Steppe, west of the Fergana valley, was proposed and some obsolete dams were destroyed in the 1870s. In 1887, the Directorate of Irrigation was established by the Governor General of the province called Turkestan at the time, to coordinate the irrigation works in the region. A large stone dam was built on Syr-Darya and the Bukhararyk canal, from Syr-Darya to Golodnaya Steppe, was constructed in 1891. A larger, 70 km long canal with an intake of about $0.3\,\mathrm{km^3/year}$, named after tsar Nicholas 1st, was inaugurated in 1896, and yet another, arterial canal to the north-eastern part of the Golodnaya Steppe was built in 1901–1913. The canal was considered the most successful irrigation project of the time in the region. The first concession for irrigation in the Bukhara region was granted in 1912.

Irrigated agriculture in the region was growing, leading to an increase of water consumption and considerable loss of water resources. A.I. Voyeykov, a well known climatologist of the time, wrote in 1882:

> ... The lower reaches [...] of the rivers feeding Aral are so dry that the existence of the Aral Sea in its present limits is a proof of our backwardness and inability to use [...] such a huge mass of flowing water and fertile silt transported by Amu and Syr. In a state where the authorities know how to properly use gifts of Nature, Aral would be receiving water [...] which is not needed for irrigation.

> Cited in Ashirbekov and Zonn (2003), translated from the Russian by P. Zavialov.

Meliorative prospecting and works continued after the socialist revolution of November, 1917. A decree entitled "Assignation of 50 million rubles for irrigation works in Turkestan" (1918) was among the first edicts signed by V.I. Lenin, the head of the new communist government of Russia. A project for a 1,500 km long trans-Caspian canal taking water from Amu-Darya was proposed in 1921. The idea of diverting a part of Amu-Darya's water resources toward the lands at the eastern coast of the Caspian Sea was moved forward at the first all-Turkmenian Congress of Soviets in 1925. In 1927, a test diversion of the water from Amu-Darya into Uzboy for irrigation of the adjacent areas was undertaken. A new 25 km long arterial canal called Kyzketken with an intake of $210 \, m^3/s$ (about $6 \, km^3/year$) from Amu-Darya was constructed by 1935. Another 110 km long canal transporting $240 \, m^3/s$ (about $7 \, km^3/year$) was intended for irrigation of the areas at the left bank of Amu-Darya. The canal named after V.I. Lenin was inaugurated in 1940.

Meanwhile, the construction of fish cannery industries started in the city of Muynak, at the southern bank of the Aral Sea, in 1933. In 1937, a secret military facility for testing biological weapons was first installed on Vozrozhdeniya Island, the largest island of the Aral Sea. The testing site was enlarged and upgraded in the late 1940s and early 1950s (Ashirbekov and Zonn, 2003).

After World War II, rapidly growing production of cotton and rice in the region implied increasing water requirements. In 1948, the Farkhad integrated water scheme, with a reservoir and a power plant, was built on Syr-Darya and used for irrigation of the entire Golodnaya Steppe. The construction of the Main Turkmen canal extending from Amu-Darya to the Caspian Sea across the Karakum Desert began in 1950 at the initiative of I.V. Stalin, the leader of the country at the time. The first portion of the Karakum canal was constructed by the late 1950s. The giant 1,300 km long canal (completed in the 1980s) supplies water to many regions of Turkmenistan. The canal was built on sands without sufficient sealing and is, therefore, subject to high seepage losses (e.g., Froebrich and Kayumov, 2004). The rate of water abstraction by the Karakum canal is estimated as $8–14 \, km^3/year$ (e.g., Hannan and O'Hara, 1998). Another major canal constructed in the 1950s is the Amu–Bukhara canal which has diverted up to $10 \, km^3$ of water annually (Ashirbekov and Zonn, 2003).

By the end of the 1950s, the water diversions in the Aral Sea basin had exceeded critical thresholds, and the progressive shallowing of the lake started in 1961. Nonetheless, the irrigation activities continued to build up. A number of reservoirs and canals diverting water from Amu-Darya and Syr-Darya, such as the South Golodnaya Steppe canal withdrawing up to $11 \, km^3/year$, the Karshi canal consuming about $5 \, km^3/year$, and others, have been constructed in subsequent years. The area of irrigated lands has increased almost twofold from $41,000 \, km^2$ in 1960 to $74,000 \, km^2$ in 1990 (Ashirbekov and Zonn, 2003). The region accounted for about 95% of cotton, 40% of rice, and about 30% of fruits and vegetables produced in the USSR. The construction works on new integrated irrigation systems had mainly ceased by the 1990s, when the lake level had dropped by about 16 m and the northernmost portion of the lake ("Small Sea") detached from the main body.

In the 1990s and 2000s, following the decay of the former USSR and the establishment of the newly independent states in Central Asia, the Aral crisis has attracted broad attention at the international level. The International Coordination Water Commission in Central Asia (ICWC) and the International Aral Salvation Foundation have been established since the early 1990s. A number of projects aimed at improving the water management policy in the region and easing the consequences of the crisis have been implemented with varying degrees of success. The efforts have been ensured by diversified support from a number of international and national organizations and entities. The related issues, which are beyond the scope of this book, have been broadly discussed elsewhere in recent literature. Other significant landmarks of the time were the closure of the Vozrozhdeniya weapon-testing site in 1992 and the subsequent decontamination of the island, and the beginning of extensive geological prospecting for petroleum in the Aral Sea region.

The ongoing efforts to ease the consequences of the ecological disaster imply the necessity for better understanding of the hydrological state of the Sea and physical and chemical processes taking place in its waters. From the physical oceanographic viewpoint, the present lake differs significantly from the Aral Sea during its pre-desiccation state, which has been extensively explored.

One of the first geographic expeditions to Khiva and the region east of the Caspian Sea was organized by A. Bekovich-Cherkasskiy on the initiative of the tsar Peter the Great as early as 1715. The first geodesic mapping of the Aral Sea shore is ascribed to another Russian survey of 1731 (Ashirbekov and Zonn, 2003). A number of other notable expeditions to the area, such as those headed by I. Muravin and D. Gladyshev (1741), E.A. Eversman (1820s), Colonel F.F. Berg (1823, 1825–1826), and G.I. Danilevskiy and F.I. Baziner (early 1840s), which took place in the subsequent century, yielded improved maps and important new data about the geography and climate of the Aral Sea region (Yanshin and Goldenberg, 1963; Bortnik and Chistyaeva, 1990). However, the first detailed description of Aral's water body was given in 1848–1849 by Commander A.I. Butakov and his crew of 27 who sailed in the Aral Sea on the schooner *Konstantin* built in Russia and delivered to fort Raim at the mouth of Syr-Darya. The information collected during the survey encompassed accurate determination of the shoreline coordinates, bathymetry mapping, and hydrographic observations. The expedition discovered the islands of the Aral Sea, including the largest one, initially named after tsar Nicholas 1st and presently known as Vozrozhdeniya. Butakov's map of the Aral Sea is shown in Figure 1.2.

After the Khiva khanate had became a part of the Russian customs territory in the 1870s, geographic research and exploration in the Aral Sea region intensified. In 1874, A.A. Tillo deployed the first geodesic benchmark in the region and performed accurate geodesic leveling (Tillo, 1877). The first determination of the salt content in the Aral Sea water was published in 1870, and that of the salt composition in 1872, based on a sample collected by Captain C. Sharngorst on his way from St. Petersburg to Bukhara. Later he wrote:

... While men were changing the horses at the station Ak–Dzhulpas, I got into the lake up to my knees and filled two Champagne bottles with the water.

Cited in Blinov (1956), translated from the Russian by P. Zavialov.

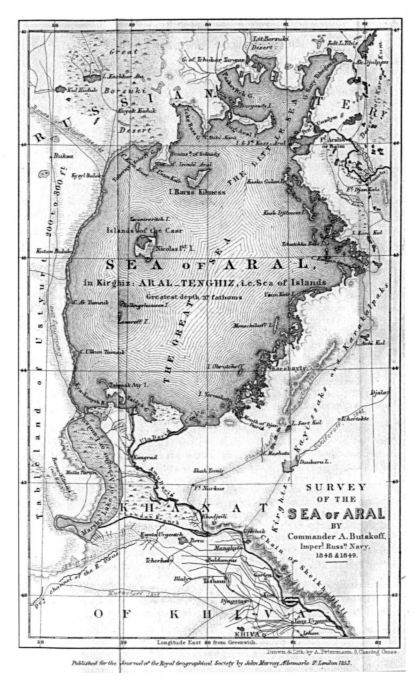

Figure 1.2. The first detailed map of the Aral Sea published in 1853 by Commander A. Butakov, Imperial Russian Navy, based on his survey of 1848–1849.

From http://www.lib.utexas.edu/maps/historical, October 2004.

Despite the simplicity of this sampling procedure, chemical analysis of the water performed in Tashkent immediately revealed the peculiar salt composition of the Aral Sea.

A large-scale interdisciplinary survey of the Aral Sea headed by L.S. Berg in 1900–1903 yielded a wealth of new hydrological and geographic information published later in a fundamental monograph (Berg, 1908). L.S. Berg was also the first to organize direct instrumental measurements of the lake level variability in 1900. Since 1911, the level changes have been continuously recorded at a meteoro-logical station in the city of Aralsk, at the northern extremity of the Sea. Biological and fishery related studies in the Aral Sea have rapidly progressed since the 1920s, especially after the establishment of the Aral Scientific Fishery Station in 1929 (e.g., Ashirbekov and Zonn, 2003). In particular, experiments on acclimatizing fishes from other basins have been undertaken (see, e.g., a review by Aladin et al. (2004)). An important monograph describing fishes of the Aral Sea as well as the regime of the lake's waters was published by Nikolskiy (1940).

A network of permanent hydrometeorological stations around Aral's shore and on the islands has been expanding since the 1930s. Overall, the network encompassed as many as 11 meteostations, up to 9 of which were in simultaneous operation in the 1950s and 1960s (Bortnik and Chistyaeva, 1990). A vast program of hydrographic, chemical, biological, and geological research has been implemented since the 1950s. A new handbook of sailing directions for the Aral Sea was published (*Sailing Directions for the Aral Sea*, 1963). Standard observations from research vessels on an extensive grid of oceanographic stations and transects have been organized on a routine basis several times a year, in addition to continuous measurements at the coastal sites, aircraft observations, etc. For example, chemical sampling was regularly performed at about 48 stations, as reported by Blinov (1956), while bottom sediments have been collected at 127 locations all over the Sea (Rubanov et al., 1987). Aral's circulation, water and salt budgets, physical and chemical states, and biological productivity, have been investigated by means of data analyses as well as laboratory and modeling studies. The research was conducted by specialists from the State Oceanographic Institute, Kazakhstan Institute of Fisheries, Hydro-meteorological Services of Kazakhstan and Uzbekistan, institutes of the Academy of Sciences, and many other organizations of the former USSR. By the 1980s, a wealth of information had been obtained thereby, and the Aral Sea was probably one of the most well sampled inland water bodies of the planet. The newly obtained data were generalized and published in a number of classical literature sources. However, as discussed in the Introduction, the field research in the lake significantly reduced at the advanced stages of desiccation, in the 1990s, following the collapse of the USSR.

2

The immediate past: A summary of the pre-desiccation state

In this chapter, we give a summary of the physical state of the Aral Sea prior to the shallowing onset in 1961. Although we use the present tense throughout the chapter for convenience, the figures and the data presented in this chapter refer to the period from 1911, when instrumental measurements in the Aral Sea began on a more or less regular basis, through to 1960. The drastic changes which have occurred after 1960 are addressed in Chapter 3.

In 1960, the lake surface level was slightly over 53 m above the mean ocean level. It should always be kept in mind, however, that the Aral Sea level has been subject to a considerable interannual variability even before the desiccation began. The interannual changes of the lake level over the period 1911 through to 1960 were as large as ±0.5 m (Lvov, 1966), and, according to most reconstructions, the interannual and decadal range of the lake level was over 2 m in the 19th century (Berg, 1908; Lvov, 1959). This variability was also accompanied by a seasonal cycle with an amplitude up to 30 cm (e.g., Lymarev, 1967; Lvov, 1970a), locally, very strong wind-controlled surges (up to ±180 cm (Bortnik and Chistyaeva, 1990)), and other short-scale variability. Nonetheless, the 53-m level is traditionally considered to be characteristic of the "normal", pre-desiccation state of the Aral Sea, therefore, most of the figures given in this chapter correspond to this value.

2.1 PHYSICAL GEOGRAPHIC SETTINGS

In 1960, the Aral Sea was the fourth largest inland water body on earth, after the Caspian Sea, Lake Superior, and Lake Victoria. The Sea was located between 43°24′N and 46°53′N, and 58°12′E and 61°59′E, with its longitudinal axis slightly tilted in the NE–SW direction. The surface area of the Sea was over 66,000 km², with the maximum length of 492 km and maximum width of 290 km (Rubanov et al., 1987). The total extent of the shoreline was nearly 3,000 km. The mean depth of the lake was about 16 m. Early studies have reported maximum depths of up to 69 m (e.g., Butakov, 1853; Fortunatov and Sergiyenko, 1950; Betger, 1953) which was later replicated in a number of publications and even navigation maps. More recent investigators questioned the existence of depths exceeding 63 m (e.g., Rubanov et al., 1987). There were more than 1,100 small and large islands in the Aral Sea occupying a total area of over 2,200 km^2. As mentioned in Chapter 1, it was the abundance of islands that gave the name to the Sea—in Turkic, the word "aral" means island.

In geodynamical terms, the Aral Sea is located in the northern part of a broad continental platform, the Turan Plate. The southern extremity of the plate coincides with the reverse-dextral Ashgabat fault. The region is subject to earthquake activity. The Sea occupies the lowest part of a large erosion–tectonic depression, extending about 800 km in the NE–SW direction. The Earth's crust thickness in the Aral Sea region is 35–40 km (Volvovskiy et al., 1966). The granite layer thickness is 12–18 km, and that of the basaltic layer is 16–18 km (Inogamov et al., 1979).

A schematic map of the Aral Sea corresponding to 1960 is shown in Figure 2.1. The Sea is fed by two principal rivers of Central Asia, the Amu-Darya and Syr-Darya with an average combined annual runoff of about 56 km^3 (Bortnik and Chistyaeva, 1990). Amu-Darya merges into the southernmost portion of the Sea, while Syr-Darya feeds its northern part. The Amu-Darya runoffs are typically twice as large as those from Syr-Darya and the catchment basin area of the Aral Sea is almost 3 million km^3. The population in the area is above 20 million. The Sea belongs to Uzbekistan (southern part) and Kazakhstan (northern part), and these two republics, which belonged to the former Soviet Union, have been independent states since 1991.

The western bank of the Sea is formed by the abrupt cliff of the Ustyurt Plateau confined to the west by the Caspian Sea. The Ustyurt cliff towering up to 190 m above the Aral Sea surface is often referred to by local people as the *chink* [chi^ngk], which can be approximately translated as "hill top", and this idiomatic term is also not uncommon in the Russian scientific literature. The western shore is weakly indented. To the south, the Aral Sea is bordered by a vast area formed by alluvia of the ancient and contemporary Amu-Darya deltas and, farther south, sand barkhans of the Zaunguz Karakum Desert. The most notable features of the southern shore are the broad Adzhibay, Dzhiltyrbas, and Muynak Bays. On the east, the Aral Sea shore is adjacent to the Kyzylkum Desert, a sandy plain crossed by dry ancient river traces of the Syr-Darya and Amu-Darya (Rafikov and Tetyukhin, 1981). The eastern shore is strongly indented by several bays and innumerous (>600) small sand islands forming the Akpetkinskiy Archipelago in the south-eastern part of the Sea. The northern extremity of the Sea is bordered by the Barsuki and Priaral Karakum Deserts. The northern bank is rather high and steep, and broken by a number of large bays such as the Butakov, Shevchenko, and Tschebas Bays.

The bottom topography of the Aral Sea is shown in Figure 2.2. The northernmost portion of the Sea known as the Small Sea and partly isolated from the Large Sea by Kokaral Island is connected to the main part of the lake through the

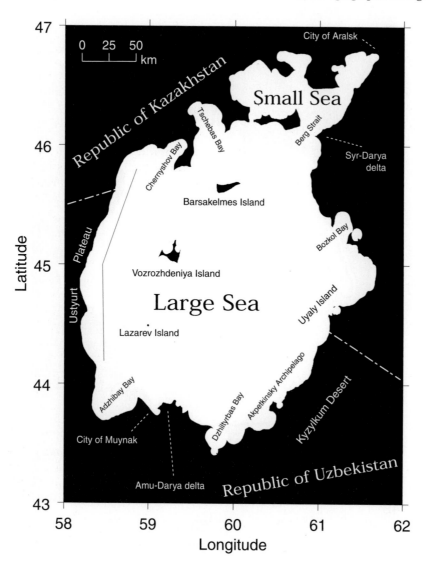

Figure 2.1. Map of the Aral Sea in 1960 and principal toponyms. The fine line in the western part of the Sea indicates the location of hydrographic sections referred to in the text.

15 km wide and 12 m deep Berg Strait east of the island and the narrow and shallow Auzy–Kokaral Strait west of it. In 1960, the maximum depth in the Small Sea was slightly below 30 m and the total volume of the Small Sea was 80 km^3 (Bortnik and Chistyaeva, 1990). In turn, the Large Sea is naturally divided into two basins, separated by an underwater tectonic swell known as the Arkhangelskiy swell extending from the Kulandy Peninsula between Chernyshov Bay and Tschebas Bay in the north to the Muynak Peninsula in the south. The highest areas of this

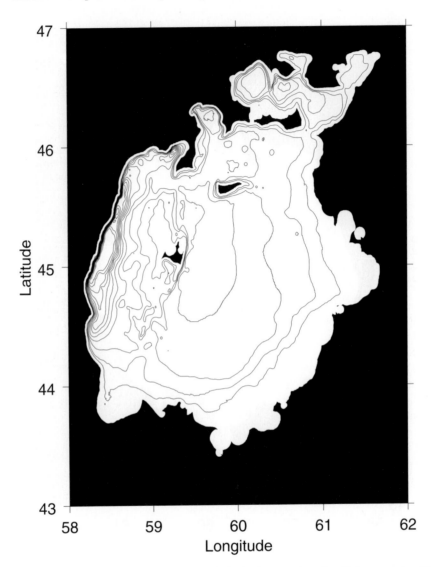

Figure 2.2. Aral Sea bottom topography. The isobaths are plotted with increments of 5 m.

shallow belt are the Vozrozhdeniya, Lazarev, and Komsomolskiy Islands. The western part of the Large Sea is a relatively deep (>60 m) trench with an abrupt bottom slope at the western shore and a more gently sloped bottom at the eastern side. The eastern basin is a large, relatively shallow (< 30 m) flat-bottomed hollow. In the northern part of the basin, there is a large island called Barsakelmes surrounded by steep bottom slopes.

In most parts of the lake's area, the bottom is constituted by silts and clayey silts.

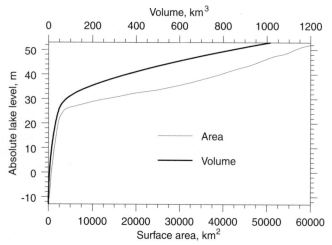

Figure 2.3. Hypsometric relations for the Aral Sea computed from detailed bottom topography shown in Figure 2.2.

Firm floors can only be seen at some isolated locations near the western coast. Some coastal regions and some areas on the Arkhangelskiy swell are sandy.

According to Kosarev (1975), the oceanographic regionalization of the Aral Sea includes 5 areas with different physical characteristics, namely: (1) the northern region (i.e., Small Sea); (2) the eastern and south-eastern region; (3) the southern region adjacent to the Amu-Darya delta; (4) the central part of the Sea; and (5) the western deep region.

For modeling and many applied purposes, it is often necessary to know the hypsometric relations of the lake (i.e., the relations interconnecting the lake surface level, the lake area, and the lake volume). The corresponding curves were obtained by directly integrating the bottom topography map at a spatial resolution of about 1 km. The relations are shown in Figure 2.3 (see Chapter 4 for a more detailed discussion). Similar curves were obtained by Mikhailov et al. (2001) using a different technique (see also Stanev et al. (2004)).

2.2 BACKGROUND METEOROLOGY AND CLIMATOLOGY

Climate in the Aral Sea region is subtropical and continental. The annual mean solar irradiance is about 185 W/m^2 (Bortnik and Chistyaeva, 1990). The Sea is located in an arid zone, and the influence of the surrounding deserts on climatic conditions is generally stronger than the moderating effect from the Aral Sea as a large water body. Such a moderation is believed to be confined to a narrow (~100 km) belt directly adjacent to the lake shore (e.g., Zhitomirskaya, 1964). Summertime conditions in the region are typically characterized by clear sky and high air temperature. Lower temperature episodes can be provoked by atmospheric intrusions from the north or north-west. In winter, the region is under the peripheral influence from the Siberian high-pressure system, which often results in clear and dry weather. At the same time, western intrusions can be accompanied by intense precipitation events.

2.2.1   Air temperature

The air temperature in the Aral Sea region generally increases from the north to the south. The lowest air temperatures, about $-6°C$ over the southern portion of the Sea and below $-12°C$ in the north of the region, on average, are observed in January (Figure 2.4). In summer, the air temperature over the Sea is typically rather uniform spatially; maximum temperatures, 25–27°C on average, are characteristic for July.

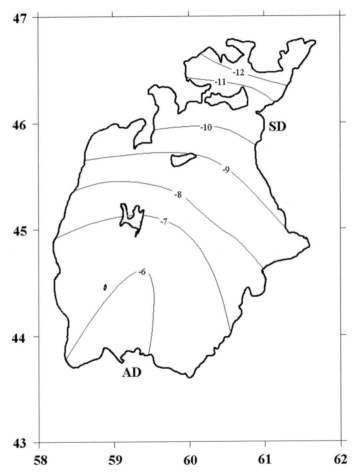

Figure 2.4. Historical average air temperature over the Aral Sea in January. The coordinate axes are latitude and longitude. Hereinafter, AD and SD indicate the respective deltas of Amu-Darya and Syr-Darya.

Redrawn from Bortnik and Chistyaeva (1990).

Table 2.1. Monthly mean air temperature (°C) in Aralsk and Muynak (1951–1960) for different months. Hereinafter, I is January, XII is December, etc.
Adapted from Bortnik and Chistyaeva (1990).

Month	I	II	III	IV	V	VI	VII	VIII	IX	X	XI	XII
Aralsk	−12.8	−10.9	−4.5	8.6	17.3	23.3	25.6	24.1	17.8	8.0	−2.9	−8.7
Muynak	−5.4	−4.7	−0.7	8.5	17.3	23.1	25.7	25.0	19.6	11.3	2.5	−2.9

The mean summer temperatures of the air over the Aral Sea are lower than those of the air over the surrounding deserts by up to 5°C.

The monthly mean air temperatures at the northern and southern extremities of the Sea for 1951–1960 are given in Table 2.1.

2.2.2   Humidity and precipitation

Relative humidity of air in the Aral Sea region in summer is 20–25% higher than that in the surrounding desert areas. In winter, the difference is largely damped. In summer months, relative humidity over the Sea usually ranges from 40–70%. In wintertime, it varies between 70 and 90%. The monthly mean relative humidity values at the northern and southern extremities of the Sea for 1951–1960 are given in Table 2.2.

Precipitation over the Aral Sea is normally small, totaling 100–140 mm/year on average, but high values of 285 mm as well as lows of only 35 mm have been documented in individual years. This interannual variability is believed to have a large spatial scale nature and be caused by processes affecting all Central Asia. At the annual scale, the maxima of precipitation are observed in spring (March–April) and autumn, and the minimum corresponds to summer (August). Overall, the characteristic number of rainy days is 30–45 per year, with snowfall days being 12–30 per year (Bortnik and Chistyaeva, 1990).

Precipitation over the Sea generally tends to increase northward (Figure 2.5). The monthly precipitation sums at the northern and southern extremities of the Sea for 1951–1960 are given in Table 2.3.

2.2.3   Winds

The regional scale atmospheric circulation in the Aral Sea is controlled to a large extent by the Siberian anticyclonic system whose center is located north-east of

Table 2.2. Monthly mean relative humidity (%) in Aralsk and Muynak (1951–1960).
Adapted from Bortnik and Chistyaeva (1990).

Month	I	II	III	IV	V	VI	VII	VIII	IX	X	XI	XII
Aralsk	81	81	78	59	49	42	44	45	46	58	73	80
Muynak	84	82	79	70	62	59	61	62	63	68	75	82

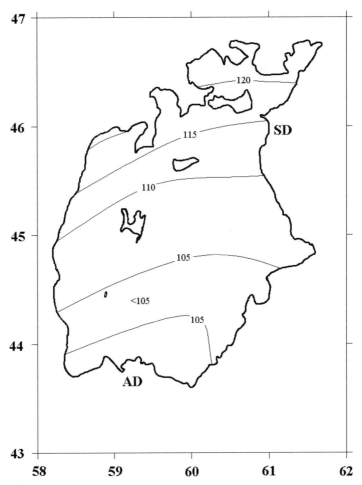

Figure 2.5. Historical average precipitation sums over the Aral Sea (mm/year). The coordinate axes are latitude and longitude.

Redrawn from Bortnik and Chistyaeva (1990).

Table 2.3. Mean monthly precipitation sums (mm) in Aralsk and Muynak (1951–1960).

Adapted from Bortnik and Chistyaeva (1990).

Month	I	II	III	IV	V	VI	VII	VIII	IX	X	XI	XII
Aralsk	8	9	16	10	15	7	16	12	4	15	17	15
Muynak	9	13	11	8	6	6	6	8	3	9	9	8

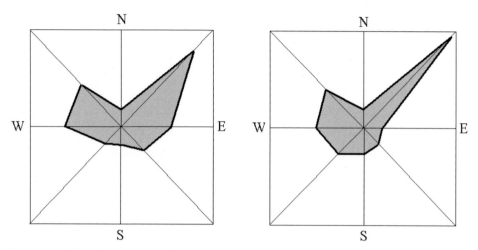

Figure 2.6. Historical average wind roses for the Aral Sea region (Aktumsyk meteorological station, western shore of the Sea, Ustyurt Plateau, day-time winds). January (*left panel*) and July (*right panel*).

Adapted from Bortnik and Chistyaeva (1990).

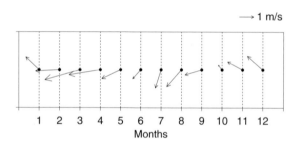

Figure 2.7. Monthly mean winds for the Aral Sea region (climatic vector averages).

Central Asia in winter and north-west of it in the summer. Consequently, the dominant winds are from the north, north-east, and north-west. The climatic wind roses typical for the Aral Sea region are exhibited in Figure 2.6. Overall, the most frequent winds are north-easterlies whose repetitivity is over 30%. The vector-averaged climatic monthly winds have been calculated from the well-known NCAR/NCEP (National Center for Atmospheric Research/National Center for Environmental Prediction, USA) reanalysis data (e.g., Kalnay et al., 1996) and are shown in Figure 2.7. The data were interpolated onto the point 45°N, 60°E in the central part of the Sea. Scalar averaging yields characteristic mean wind speeds between 3 and 7 m/s (Bortnik and Chistyaeva, 1990). Strong winds are more common at the western shore of the Sea where they occur on about 50 days per year. Wind speed values as high as 30 m/s have been documented in this region in wintertime.

Table 2.4. Average number of cloudy days per month in Aralsk and Muynak (1951–1960).
Adapted from Bortnik and Chistyaeva (1990).

Month	I	II	III	IV	V	VI	VII	VIII	IX	X	XI	XII
Aralsk	11	10	12	8	5	4	4	2	3	7	10	11
Muynak	13	11	8	8	2	2	2	1	1	4	10	13

Breeze circulation is well pronounced in summer near the shores. The diurnal amplitude of the wind speed in the cross-shore direction sometimes exceeds 3 m/s.

Strong and moderate winds often result in dust storms or near-ground "dust blizzards" which occur on up to 100 days per year and take place all over the region, but are most frequent in the area adjacent to the northern shore of the Sea. The dust storms are particularly favored by easterly and north-easterly winds which account for over 60% of such events (Grigoriev and Lipatov, 1982).

2.2.4   Cloudiness

Generally, cloudy days are not very frequent in the Aral Sea region. The number of such days per year is 60–90 (Zhitomirskaya, 1964). Maximum cloudiness is usually observed in the northern part of the Sea in winter. In summer and autumn, local convective clouds can be observed above the lake surface.

The average numbers of cloudy days per month at the northern and southern extremities of the Sea for 1951–1960 are given in Table 2.4.

2.3   WATER BUDGET COMPONENTS

2.3.1   River discharges

The net water resources of the rivers feeding the Aral Sea are estimated at about 70 km^3/year for Amu-Darya and 35 km^3/year for Syr-Darya, representing a long-term average (Asarin, 1973). A large portion of the runoff, however, is diverted for irrigation. The area of irrigated lands in the region has increased almost threefold in the past century. The total irretrievable withdrawal of water resources for irrigation is estimated at approximately 30 km^3/year before 1950 (Asarin, 1973; Volftsun and Sumarokova, 1985; Rubinova, 1987), and over 40 km^3/year for 1951–1960 (Bortnik and Chistyaeva, 1990). In addition to these anthropogenic losses, there is also a considerable water retention because of evaporation and transpiration by vegetation, especially in the delta areas. Available quantitative estimates for the delta retention strongly vary from essentially zero to 6–10 km^3/year for Amu-Darya (e.g., Samoilenko, 1955b; Rogov et al., 1968; Golubtsov and Morozova, 1972; Asarin, 1973) and 0.3–1.5 km^3/year for Syr-Darya (e.g., Zaikov, 1946; Korganov, 1969; Lvov et al., 1970; Simonov and Goptarev, 1972; Asarin, 1973; Shults, 1975). After these deductions, the contribution of continental discharge into the Aral Sea's water

Table 2.5. Summary of water budget components (except ground-water discharges) before the desiccation onset. R is the annual river runoff, P is the annual precipitation, and E is the annual evaporation from the lake surface. The first line corresponds to the entire period of observations, the second one highlights the last pre-desiccation decade.

From Bortnik and Chsityaeva (1990).

Period	R (km^3)	P (km^3)	E (km^3)
1911–1960	56.0	9.1	66.1
1951–1960	58.4	9.2	66.0

budget totaled to $56\,\mathrm{km}^3$/year on average over the period 1911–1960 (Table 2.5). In individual years, the inflow ranged from 40–$69\,\mathrm{km}^3$ (Bortnik and Chistyaeva, 1990).

2.3.2 Precipitation

Spatial and temporal variability of atmospheric precipitation over the Aral Sea was discussed in Section 2.2.2. The estimates of the total contribution from precipitation in the water budget of the Sea available in the literature are somewhat diverging. According to Samoilenko (1955b), the average input from precipitation is 5–$6\,\mathrm{km}^3$/year, or about 10% of the river discharges, which appears consistent with some earlier estimates (Zaikov, 1946; see also Blinov, 1956). On the other hand, Bortnik and Chistyaeva (1990) reported the mean value of $9.1\,\mathrm{km}^3$/year for 1911–1960, with the precipitation input ranging from 4.4–$15.5\,\mathrm{km}^3$ in individual years. Such a divergence of the estimates could be explained if there were a significant increase of precipitation in the 1950s, but the historical precipitation record does not seem to confirm this (see Table 2.5). However, the figures presented by Bortnik and Chistyaeva (1990) are likely to be more accurate as they used higher quality data from a larger number of meteorological stations.

2.3.3 Evaporation

Because very little, if any, direct measurements of evaporation from the Aral Sea had been made, the evaporation rates were usually calculated through empirical formulas specifically tuned for the Aral Sea conditions (Zaikov, 1946; Samoilenko, 1955b; Shults and Shalatova, 1964; Lvov et al., 1970; Simonov and Goptarev, 1972). Perhaps the most well known of such formulas is the relation proposed by Goptarev and Panin (1970) for obtaining the evaporation at the monthly temporal scale:

$$E = \kappa(q_s - q_z)U_z \tag{2.1}$$

where E is the evaporation rate (mm/month); q_s is the partial pressure (hPa) of saturated water vapor at the sea surface for given water temperature and salinity; q_z is the partial pressure (hPa) of the water vapor at some standard height z above the surface (e.g., $z = 2\,\mathrm{m}$); U_z is the wind speed (m/s) at the height z above the surface; and κ is a variable coefficient. Of course, this relation has the form of common "bulk formulas" routinely used in physical oceanography (e.g., Gill, 1982), but the parameterization of κ is adapted for the Aral Sea. According to Goptarev and Panin (1970);

$$\kappa = 327.5\{\ln^* z - \ln^* z_0\}^{-2} \tag{2.2}$$

where z_0 is the roughness parameter assumed to be equal to $0.0006\,\mathrm{m}$ for the Aral Sea, and $\ln^*$ is the so-called factorial logarithm:

$$\ln^*(z) = \ln(z) + \sum_{n=1}^{\infty} \frac{(\alpha z)^n}{n \cdot n!} \tag{2.3}$$

In the latter formula, α is a constant multiplier characterizing temperature stratification in the boundary layer of the atmosphere.

The scene-specific function κ was tabulated by Goptarev and Panin (1970) for a variety of air temperature profiles, and Eq. (2.1) was then used as a basis for most of the practical estimates of evaporation from the Aral Sea surface.

The figures obtained thereby were summarized by Bortnik and Chistyaeva (1990). The average net evaporation from the Aral Sea for 1911–1960 totals $100.0\,\mathrm{cm/year}$, or, in volumetric terms, $56.0\,\mathrm{km^3/year}$ (Table 2.5). We note, however, that Zaikov (1946) suggested a smaller value, namely, about $47\,\mathrm{km^3}$. The annual evaporation ranged from 44–$68\,\mathrm{km^3}$ in individual years, depending on summer air and water temperatures, and also on the duration of icy periods and the severity of winters. The evaporation rate is also subject to a remarkable seasonal modulation: the maximum values typically observed in August (except in coastal and shallow areas where the maximum evaporation usually occurs in July) exceed those characteristic for February by a factor of at least 10.

2.3.4 Groundwater exchanges

No reliable quantitative data about this component has been reported. Based on indirect estimates, most investigators agree that the contribution from the groundwater exchanges is no larger than several tenths of one $\mathrm{km^3/year}$ (e.g., Zaikov, 1952; Samoilenko, 1955b; Lvov, 1970b; Simonov and Goptarev, 1972; Asarin, 1973; Glazovskiy, 1976; Akhmedsafin, et al., 1983), although some authors argued that it could be up to $3.5\,\mathrm{km^3/year}$ (Chernenko, 1970, 1972). The upper bound for the infiltration of seawater through the bottom is estimated at $0.15\,\mathrm{km^3/year}$ (Rogov et al., 1968; Chernenko, 1972). Hence, the groundwater exchanges were normally neglected in the pre-desiccation Aral Sea water budget.

The average annual water budget components are summarized in Table 2.5.

2.4   SEA SURFACE TEMPERATURE (SST) AND ICE REGIME

2.4.1   SST variability

The overall, annually averaged SST is about $10.2°C$ for the northern part of the Sea and about $11.6°C$ for the southern portion (Kosarev, 1975). In the seasonal cycle, the surface temperature varies from $-0.5°C$ (February) to $24–25°C$ (July–August). This seasonal variability is evident at all depth levels, even in the bottom part of the western trench, where the annual range is about $3°C$. The phase delay of the seasonal temperature cycle increases with the depth by about 15 days per 10 m, so the annual temperature maximum $(3.5°C)$ in the deepest layers of the western basin occurs in November–December, followed by the annual minimum $(0°C)$ in May–June. An example of the SST statistics for different months is given in Table 2.6.

The water temperature in the Aral Sea, as a shallow water body, is also subject to strong variability at short temporal scales, associated with local meteorological forcing at the surface and, frequently, wind-induced upwellings of colder deep waters. The latter phenomenon is especially characteristic for the western steep slope, where sudden temperature drops by up to $13°C$ (!) over the period of only a few hours have been reported (Bortnik and Chistyaeva, 1990; Kosarev, 1975).

The diurnal cycle of SST in the Aral Sea is usually strong. The diurnal SST range can be up to $2°C$ over deep areas and over $3°C$ in the coastal zone, especially near the river mouths. The diurnal changes decrease with depth and are traced up to a depth of about 20 m.

2.4.2   Ice regime

Because the salt composition of the Aral Sea water is different from that of the World Ocean, the respective freezing temperatures are also slightly different. The following empirical formula has been used to calculate the Aral water freezing temperature for different salinities (Kosarev, 1975):

$$\tau = -0.086 - 0.064533\sigma_0(S) - 0.0001055\sigma_0^2(S) \tag{2.4}$$

where τ is the freezing point $(°C)$, and $\sigma_0(S)$ is the density σ_t of the water for salinity S and temperature $t = 0°C$. The corresponding values are shown in Table 2.7

Table 2.6. Basic statistics for SST $(°C)$ in different months (1949–1960). The notation σ is used for the root mean square deviation. The data were collected near Barsakelmes Island in the central part of the Sea.

Adapted from Bortnik and Chistyaeva (1990).

Month	I	II	III	IV	V	VI	VII	VIII	IX	X	XI	XII
Mean	−0.4	−0.3	0.1	4.8	12.8	19.7	23.9	24.3	20.2	13.5	6.6	1.7
σ	0.5	0.2	0.5	1.3	1.0	0.8	0.5	0.6	0.9	1.3	1.8	1.4
Max.	2.0	0.9	4.1	12.4	20.0	25.0	26.6	27.5	25.1	22.4	14.2	6.4
Min.	−0.7	−0.6	−0.6	−0.2	6.0	13.6	20.6	17.9	13.3	4.0	−0.3	−0.6

Table 2.7. Freezing temperatures for the Aral Sea and World
Ocean (°).

From Kosarev (1975).

Salinity (ppt)	Aral Sea	World Ocean
4.0	−0.23	−0.21
6.0	−0.35	−0.32
8.0	−0.46	−0.43
10.0	−0.57	−0.53
12.0	−0.68	−0.64

together with the respective data for the World Ocean. For the salinity range characteristic of the Aral Sea, the freezing temperature varies between −0.5°C and
−0.7°C.

Typically, floating ice in the Aral Sea first appears in the Small Sea and the
north-eastern part of the Large Sea in late November. At the southern extremity of
the Sea, the first ice is normally observed in mid-December. The spatial distribution
of floating ice is largely wind-controlled (e.g., ice is often accumulated in the
southern part of the Sea under the predominant north-easterly winds (Kosarev,
1975)). The deep western basin remains ice-free at least until the first decade of
January. Pack ice is first observed in mid-December at the northern shore. By
January, it covers the entire Small Sea and the eastern and southern parts of the
Large Sea. In severe winters, the Aral Sea is completely covered with ice by late
January. In mild and moderate winters, the ice cover is only partial. The spatial
extent of the ice cover and ice thickness in different months and decades are illustrated by Table 2.8. As it is seen, the maximum ice cover in February is normally
close to 90% of the Aral Sea area. The ice thickness data given in the table refer to
the northern shore of the Small Sea near Aralsk, where ice is normally the thickest
for the Aral Sea—values up to 100 cm are not uncommon in severe winters (Kosarev,
1975). The pack ice thickness generally decreases southward to 30–40 cm (on
average) in the southernmost sector of the Sea. The total duration of the icy
period varies from 70–80 days/year for the western deep basin to 120–140 days/
year for the northern and eastern parts of the Sea.

Table 2.8. Fraction of the Aral Sea area covered by ice (%) and pack ice thickness (cm) near
Aralsk in different months and decades (decade 1 refers to days 1–10, decade 2 to days 11–20,
and decade 3 to days 21–31 of the month). The data are from numerous ice surveys conducted
in 1950–1960.

From Bortnik and Chistyaeva (1990).

Month	XI	XII	I	I	I	II	II	II	III	III	III	IV	IV
Decade	2	3	1	2	3	1	2	3	1	2	3	1	2
Area	11	29	45	64	80	88	88	85	79	69	56	42	20
Thickness	25	36	47	52	56	61	65	68	69	71	68	69	43

2.5 THERMOHALINE AND DENSITY STRUCTURE

2.5.1 Temperature

One of the key physical processes determining the 3D temperature structure in the Aral Sea is convection. Thermal convection is important in autumn and winter. As well as for the World Ocean water, the maximum density temperature Θ for the Aral Sea water is higher than the freezing temperature τ. Because of different salt compositions, Θ for the Aral Sea is different from that of the World Ocean. The following empirical formula has been used to calculate the maximum density temperature for the Aral water at different salinities (e.g., Kosarev, 1975):

$$\Theta = 3.95 - 0.266\sigma_0(S) \tag{2.5}$$

where Θ is expressed in $°C$, and $\sigma_0(S)$ is the density σ_t of the water for salinity S and temperature $t = 0°C$. The values are shown in Table 2.9 together with the respective data for the World Ocean. For the salinity range characteristic of the Aral Sea, the maximum density temperature varies between $1.2°C$ and $3.1°C$.

In autumn and early winter, surface cooling results in unstable density stratification in the upper portion of the water column and, consequently, convection onset. This thermal convection progresses until the surface cools down to the maximum density temperature. Then, under continuing cooling, the upper layer becomes stably stratified again, which leads to an accelerated decrease of surface temperature. As soon as the temperature attains freezing point, the salt released in the upper layer, because of the ice formation, triggers haline convection which continues until January–February, as long as the ice grows. Ice melting in spring restores stable haline stratification, but thermal convection is possible while the temperature at the surface is still near the maximum density temperature. In summer, haline convection occurs because of enhanced evaporation and corresponding salinization of the surface layer. This mechanism is especially pronounced in the shallow eastern part of the Large Sea and the near-shore shoals elsewhere. According to Simonov (1962), the saltier bottom waters originated from summer haline convection in the eastern basin may then slip downslope as a gravity current into the western trench, thus contributing to the formation of the western basin bottom water. Some investigators

Table 2.9. Maximum density temperatures for the Aral Sea and World Ocean ($°C$).

From Kosarev (1975).

Salinity (ppt)	Aral Sea	World Ocean
4.0	3.02	3.13
6.0	2.55	2.71
8.0	2.07	2.29
10.0	1.60	1.86
12.0	1.13	1.43

argued that the eastern basin waters may also be advected into the western trench in late winter, following the winter convection in the eastern basin (Kosarev, 1975).

Thus, 3 relevant basic types of convection can be identified, namely: (1) winter thermal convection; (2) winter haline convection associated with ice growth; and (3) summer haline convection linked with evaporation. These respective types are sometimes called the subpolar type, polar type, and subtropical type (e.g., Zubov, 1947). As emphasized by Kosarev (1975), despite the relatively small spatial extent of the Aral Sea, all 3 types are well manifested there.

In winter, most parts of the water body (except the bottom layer of the western trench in some years) are practically isothermal at 0–1°C because of intense convective mixing. Thermal convection typically reaches depths of about 35 m, so the areas where the total depth is shallower are fully mixed every winter. Deeper layers, namely those in the western trench, can only be affected by haline convection if the newly formed ice in the open sea is 80 cm thick or more, which can happen in severe winters (Kosarev, 1975). Top-to-bottom thermal homogeneity of the water column in the western trench is typically observed in one winter out of 3–4.

In spring, the first areas to warm up following the increase of insolation are the shallow eastern and southern parts of the Sea (Figure 2.8). In the southern part adjacent to the Amu-Darya mouth where the water column stratification is most stable because of surface freshening by river discharge, the surplus of solar heat is mainly distributed within a narrow subsurface layer and, therefore, the spring temperature increase is the largest. This effect is also evident but less pronounced near the Syr-Darya delta. By May, the SST near the southern shore is above 16°C. In the Small Sea and the northern part of the Large Sea, the surface temperature varies between 11°C and 13°C. On the other hand, the deep basin is more inertial thermally, and its heat content deficit is manifested by lower spring SSTs (9–10°C) in the western part of the Sea.

In summer, the SST distribution over the Sea is almost uniform (Figure 2.9) and the surface temperature, typically, varies between 23°C and 26°C. The isotherms are oriented zonally, suggesting that the temperature field is mainly controlled by insolation, while the maximum temperature is observed in the near-shore areas in the southern and south-eastern parts of the Sea.

In autumn, cooling starts from the shallow eastern and southern areas. At the eastern extremity of the Sea, the SST is already below 10°C by October (Figure 2.10), and the horizontal gradients are quite steep. In contrast, the waters in the western basin and the central part of the Sea are still warm at over 15°C.

Vertical sections of temperature along the axis of the deep western trench are depicted in Figures 2.11–2.13 (the section location is indicated in Figure 2.1). The deepest layers of the Sea are always occupied by very cold water (0–2°C in winter and spring, 3–5°C in summer and autumn). In spring, a very shallow upper mixed layer is followed by a relatively weak (0.2°C/m to 0.3°C/m) thermocline extending down to depths of 20–25 m, while the deeper layers are still almost isothermal. The downward temperature decrease is steeper at the southern slope which is under the influence of Amu-Darya's freshwater runoff.

In summer, the mixed layer depth is about 15 m all over the Sea, and the depth

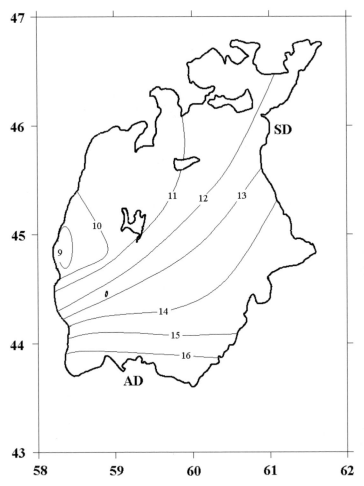

Figure 2.8. Surface temperature distribution in May (°C). The coordinate axes are latitude and longitude.

Redrawn from Kosarev (1975).

interval of 15–30 m is occupied by a well-developed thermocline where the vertical temperature gradient is up to 1°C/m and higher. Below it, there is a more homogeneous layer at 3–5°C. The lowest summer temperatures at the bottom are associated with the southern extremity of the deepest region.

In autumn, the mixed layer is best developed and the thermocline is the deepest with its upper limit located at 20–25 m, but the temperature gradient in the thermocline is typically smaller than that in the summer by approximately a factor of 2. Another notable autumn feature is an "island" of warmer surface water in the central part of the section (see also Figure 2.10). The vertical extent of this warm structure associated with the deepest portion of the basin is about 10 m.

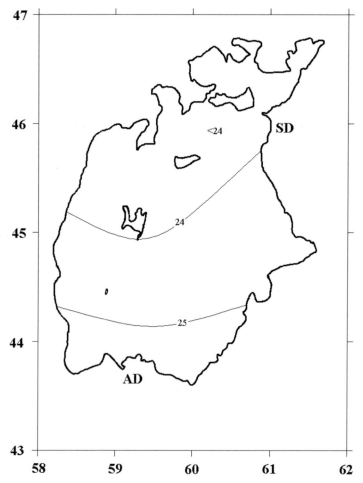

Figure 2.9. Surface temperature distribution in August (°C). The coordinate axes are latitude and longitude.

Redrawn from Kosarev (1975).

2.5.2   Salinity

The pre-desiccation Aral Sea was a brackish water body whose mean salinity was about 9.9 ppt in 1960 (Bortnik and Chistyaeva, 1990) and about 10.2 ppt (Blinov, 1956; Simonov and Goptarev, 1972) on the long-term average. The salinity field in the Aral Sea is controlled by river discharges, ice formation and melting, and evaporation and precipitation, which is reflected in its spatial structure and temporal variability. All over the Sea, salinity exhibits moderate but considerable seasonal cycling. Basic statistics for surface salinity are given in Table 2.10. These data correspond to the waters near Barsakelmes Island surrounded by steep bottom

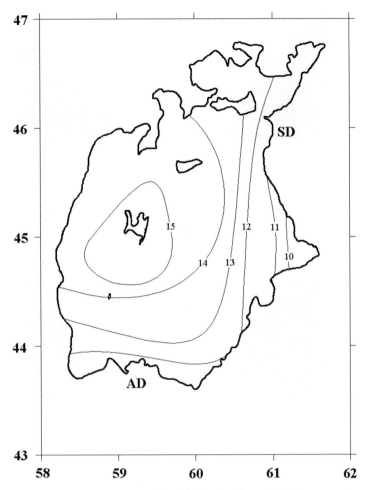

Figure 2.10. Surface temperature distribution in October (°C). The coordinate axes are latitude and longitude.

Redrawn from Kosarev (1975).

slopes in the central part of the Sea. As seen from the table, the seasonal salinity range at this location is about 1.8 ppt. According to Kosarev (1975), seasonal salinity changes in the open sea are smaller. The underlying interannual variability range is normally within 1 ppt, but it can be up to 4 ppt for spring months, which is associated with varying ice conditions and also intensity and timing of maximum fluvial discharges in individual years. The maximum seasonal range is observed in the Small Sea where the ice conditions in winter are most severe.

In winter and spring, the salinity distribution is rather homogeneous, both horizontally and vertically, because of the winter convective mixing of the basin. Salinity slightly increases from the south-western part of the Sea, where the influence

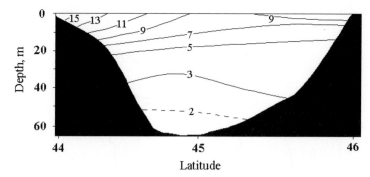

Figure 2.11. Temperature in spring (°C). Longitudinal vertical section through the western deep basin.

Redrawn from Bortnik and Chistyaeva (1990).

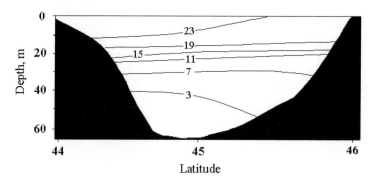

Figure 2.12. Temperature in summer (°C). Longitudinal vertical section through the western deep basin.

Redrawn from Bortnik and Chistyaeva (1990).

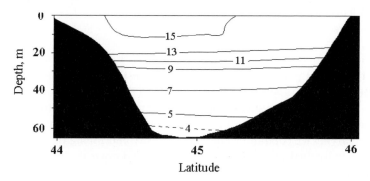

Figure 2.13. Temperature in autumn (°C). Longitudinal vertical section through the western deep basin.

Redrawn from Bortnik and Chistyaeva (1990).

Table 2.10. Basic statistics for surface salinity (ppt) in different months (1949–1960). The notation σ is used for the root mean square deviation. The data were collected near Barsakelmes Island in the central part of the Sea.

Adapted from Bortnik and Chistyaeva (1990).

Month	I	II	III	IV	V	VI	VII	VIII	IX	X	XI	XII
Mean	10.8	10.4	9.0	9.3	10.6	10.7	10.7	10.5	10.5	10.4	10.4	10.6
σ	0.2	0.7	1.3	1.5	0.3	0.4	0.3	0.3	0.3	0.2	0.2	0.4
Max.	11.2	11.2	10.7	11.0	11.0	11.3	11.3	11.2	11.0	11.7	10.8	11.2
Min.	10.4	9.6	6.5	6.5	10.3	10.2	10.3	10.0	10.1	10.0	10.0	10.0

of Amu-Darya discharge is well pronounced, to the eastern shoals. Another local minimum is seen in the area adjacent to the Syr-Darya delta and the southern part of the Small Sea (Figure 2.14).

In summer, the surface salinity distribution is similar to that in spring, but the tongue of relatively low salinity originating from the Amu-Darya mouth area is more elongated and transported farther along the western coast by anticyclonic circulation (Figure 2.15). Salinity gradients are steeper, especially in the southern and eastern regions. Because of enhanced evaporation in the shallow eastern part of the Sea, salinity increases and attains its maximum value in this area. The absolute salinity maximum ever observed in the pre-desiccation Aral Sea was registered near Uyaly Island at the eastern extremity of the lake at 16.9 ppt.

The salinity field for autumn is more homogeneous, because both river runoff and evaporation are small. The low-salinity belt along the western coast is reduced (Figure 2.16). The maximum salinity values are still associated with the eastern part of the Sea, and the minimum ones are observed in the near-delta areas. According to Bortnik and Chistyaeva (1990), the spatial inhomogeneity range in this season rarely exceeds 1.5 ppt.

Meridional vertical sections of salinity along the western deep basin are shown in Figures 2.17–2.19. The maps for spring, summer, and autumn are much alike. As expected, salinity increases downwards, but the difference between the salinity values at the surface and the bottom does not exceed 0.5 ppt (except in the area directly adjacent to the Amu-Darya mouth). The halocline which is generally rather weak is best developed in spring in the central part of the basin. The saltiest water (>10.2 ppt) occupies the layers below 30 m in the deepest part of the trench. Salinity values up to 10.5 ppt have been reported for this area in summer (Revina et al., 1970), which is thought to be associated with the downslope penetration of eastern basin waters in the bottom layer (Simonov, 1962).

2.5.3 Density

Density of the Aral Sea water is determined by its temperature and salinity. The dependence on pressure, generally taken into account for calculating the density of

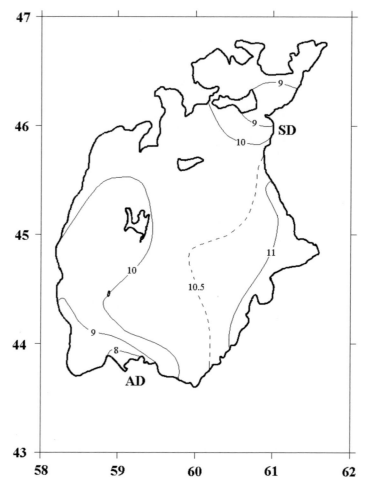

Figure 2.14. Surface salinity distribution in May (ppt). The coordinate axes are latitude and longitude.

Redrawn from Blinov (1956).

the World Ocean water, can be neglected because the Aral Sea is shallow. Because of different salt compositions, the equation of state for the Aral Sea water (i.e., the relation between density on the one hand and temperature and salinity on the other), is different from that for the World Ocean water: the Aral Sea water is about 10% heavier than the World Ocean water at the same salinity and temperature. The density relation specific for the Aral Sea has been tabulated and also expressed in the form of empirical formulas. As an example, we present here the most well-known of such expressions, namely, the Blinov formula (Blinov, 1975):

$$\sigma_{17.5} = 0.464 + 2.330Cl \tag{2.6}$$

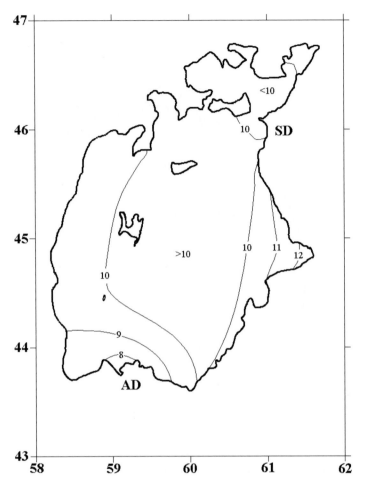

Figure 2.15. Surface salinity distribution in July (ppt). The coordinate axes are latitude and longitude.

Redrawn from Blinov (1956).

where $\sigma_{17.5}$ is the density σ_t for $t = 17.5°C$, and Cl is chlorinity (ppt), which in turn is connected with the salinity (ppt) through the relation:

$$S = 0.264 + 2.791Cl \tag{2.7}$$

For an overall spatial and long-term temporal average, the density σ_t characteristic for the Aral Sea is about $8\,\mathrm{kg/m^3}$. The seasonal cycle of density is determined by seasonal variability of temperature and, to a lesser extent, that of salinity as discussed in the preceding sections. The seasonal changes of density are summarized in Table 2.11. Typical horizontal distributions of density for different seasons are shown in Figures 2.20–2.22.

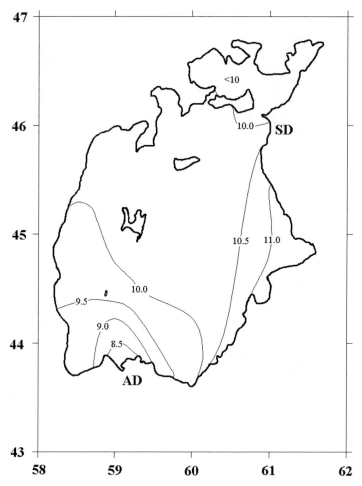

Figure 2.16. Surface salinity distribution in October (ppt). The coordinate axes are latitude and longitude.

Redrawn from Blinov (1956).

In winter, density fields are rather homogeneous at 9–10 kg/m^3 because of the ongoing convective mixing. In spring, they are mainly controlled by rapidly increasing temperatures (except in the near-delta areas, where the salinity changes can play an important role), and the density starts to decrease following the general warming trend, especially in the eastern and southern parts of the Sea.

In summer, when the SST is almost uniform over the lake area, density distribution at the surface largely follows the salinity isolines. The maximum density values of over 6 kg/m^3 are observed in the northern part of the Small Sea (Sarychaganak Bay) and in the coastal part of the eastern basin of the Large Sea where evaporation is maximum. The minimum density areas are those adjacent to river mouths where σ_t is 5 kg/m^3 or lower.

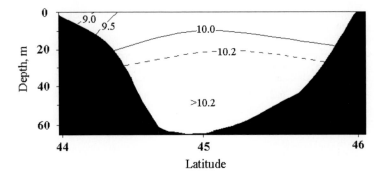

Figure 2.17. Salinity in spring (ppt). Longitudinal vertical section through the western deep basin.

Redrawn from Bortnik and Chistyaeva (1990).

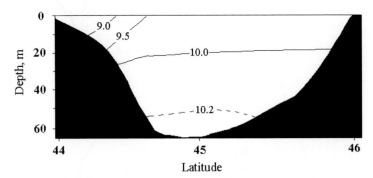

Figure 2.18. Salinity in summer (ppt). Longitudinal vertical section through the western deep basin.

Redrawn from Bortnik and Chistyaeva (1990).

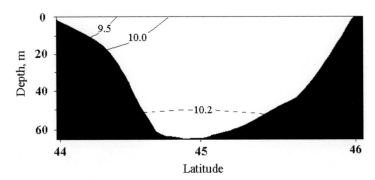

Figure 2.19. Salinity in autumn (ppt). Longitudinal vertical section through the western deep basin.

Redrawn from Bortnik and Chistyaeva (1990).

Table 2.11. Long-term average monthly density σ_t (kg/m^3) of Aral Sea water in Aralsk and Barsakelmes (1949–1960).
Adapted from Bortnik and Chistyaeva (1990).

Month	I	II	III	IV	V	VI	VII	VIII	IX	X	XI	XII
Aralsk	9.8	9.5	6.2	5.8	7.1	6.4	6.0	6.5	7.4	8.9	9.3	9.6
Barsakelmes	9.6	9.2	8.0	8.2	9.0	8.0	6.7	6.1	6.6	7.7	8.7	9.3

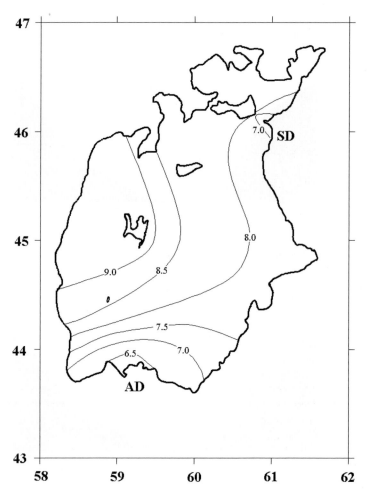

Figure 2.20. Surface density (σ_t) distribution in spring (kg/m^3). The coordinate axes are latitude and longitude.
Redrawn from Bortnik and Chistyaeva (1990).

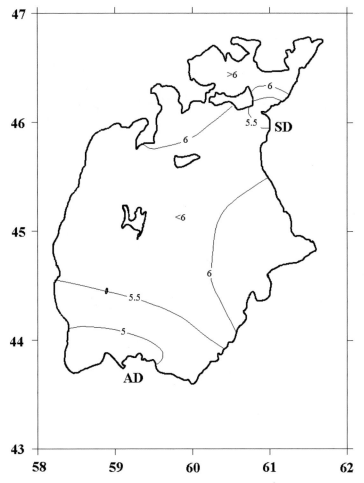

Figure 2.21. Surface density (σ_t) distribution in summer (kg/m^3). The coordinate axes are latitude and longitude.

Redrawn from Bortnik and Chistyaeva (1990).

In autumn, temperature variability regains control over the density which starts to increase in a rather uniform spatial pattern (Figure 2.22), but lighter water is still present over the deep western and central regions and in the river discharge areas.

Vertical sections of density are depicted in Figures 2.23–2.25. The density difference between the surface and the deepest layers is about 3.5 kg/m^3 in summer when the density stratification is the largest. The maximum density gradient in the picnocline is up to 0.2 kg/m^4. The main picnocline is typically located between depths of 10 and 20 m. In the other seasons, the vertical density structure is smoother and the surface-to-bottom density difference is normally below 1.5 kg/ m^3. The vertical stability of the water column in August exceeds that in May or October by almost a factor of 10.

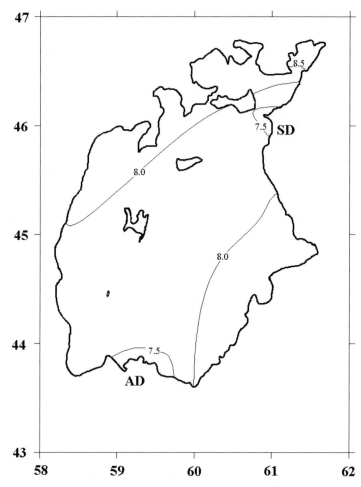

Figure 2.22. Surface density (σ_t) distribution in autumn (kg/m^3). The coordinate axes are latitude and longitude.

Redrawn from Bortnik and Chistyaeva (1990).

2.6 OPTICAL PROPERTIES OF WATER

The optical transparency of the Aral Sea water, as measured by Secchi disk, varied in individual observations in a broad range between below 1 m near the Amu-Darya and Syr-Darya mouths to above 20 m in the deep part of the Sea and in the Small Sea. The record transparency of 27 m was documented for Chernyshev Bay (Romashkin and Samoilenko, 1953). The mean values, however, are much lower: for the overall average, the transparency of the Aral Sea is about 9 m. The transparency isolines generally follow the bathymetry contours. The maximum values (12–14 m on average) are characteristic for the central and northern parts of the deep

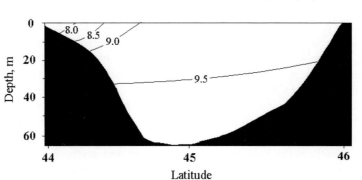

Figure 2.23. Density σ_t in spring (kg/m^3). Longitudinal vertical section through the western deep basin.

Redrawn from Bortnik and Chistyaeva (1990).

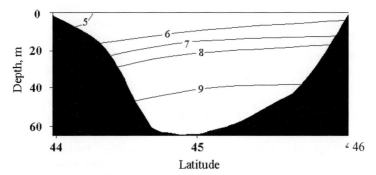

Figure 2.24. Density σ_t in summer (kg/m^3). Longitudinal vertical section through the western deep basin.

Redrawn from Bortnik and Chistyaeva (1990).

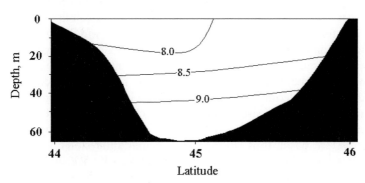

Figure 2.25. Density σ_t in autumn (kg/m^3). Longitudinal vertical section through the western deep basin.

Redrawn from Bortnik and Chistyaeva (1990).

western basin and Chernyshev Bay. Minimum transparency (2–4 m on average) is observed in the southern part of the Sea adjacent to the Amu-Darya delta. In the Small Sea, the transparency typically varies between 5 and 10 m. Seasonal variability of the transparency is rather slim, but generally, maximum values are observed in summer (Bortnik and Chistyaeva, 1990).

The dominant colors of the Aral Sea water are blue and greenish-blue (i.e., color values 4–6 in the standard color scale) (Bortnik and Chistyaeva, 1990). Only in the southernmost part of the Sea, and also in the area immediately adjacent to the Syr-Darya mouth, is the typical color bluish-green and green (7–9). Yellowish and brownish tones are occasionally seen near the river mouths in about 1% of observations.

2.7   CIRCULATION

The notion of the pre-desiccation Aral circulation is somewhat speculative, because direct current measurements were extremely few in number (Bortnik and Chistyaeva, 1990). Most of the available information was obtained through analytical and numerical modeling (Simonov, 1954; Shkudova and Kovalev, 1969; Filippov, 1970; Bortnik and Dauliteyarov, 1985). However, even early researchers noticed the most important property of the large-scale circulation in the Aral Sea, namely, its anticyclonic character (e.g., Berg, 1908; Zhdanko, 1940) under the prevailing winds. Such a predominance of anticyclonic vorticity is quite amazing since, as known, the neighboring enclosed seas of the same latitude belt (e.g., the Caspian Sea, the Black Sea, the Azov Sea) all manifest cyclonic basin-scale circulations. The opposite sign of the Aral Sea surface currents has been attributed to a combined dynamical effect of the regional winds, freshwater discharges from Amu-Darya and Syr-Darya, and rather specific bottom topography of the Aral Sea (Simonov, 1954).

A generic scheme of the mean large-scale circulation at the Aral Sea surface is shown in Figure 2.26. The Amu-Darya waters entering the Sea at its southern extremity veer north-westward (amazingly, against the Coriollis force!), which is believed to be forced by the predominant north-easterly winds together with topographic effects, and the current then spreads northward along the longitudinal axis of the deep western trench. At the same time, a weak southward transport takes place over the western slope where the depth is smaller, resulting in secondary cyclonic cells in the central westernmost coastal part of the basin and Chernyshov Bay. The mainstream current turns east between Barsakelmes Island and Vozrozhdeniya Island, and then returns to the southern part of the Sea as a broad southward flow in the eastern basin, thus closing the principal anticyclonic gyre in the Large Sea. There is a hint of a smaller cyclonic gyre between Barsakelmes Island and the Berg Strait. Therefore, there is a convergence zone in the northern part of the Large Sea north of Barsakelmes Island (Kosarev, 1975). On the other hand, under the southern winds which are relatively rare, the situation is the opposite: the main circulation gyre is anticlockwise, and there is a divergence in the northern part of

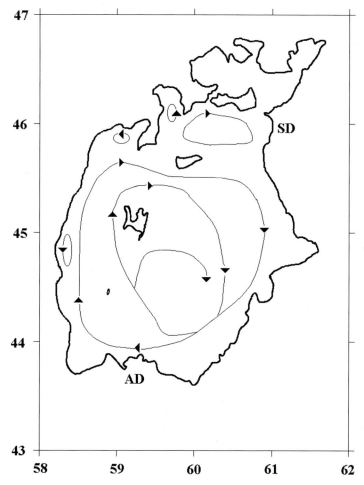

Figure 2.26. Schematic representation of the mean basin-scale Aral Sea circulation under predominant north-easterly winds.

Adapted from Rubanov et al. (1987).

the basin (Simonov, 1954). Typical surface velocity values are 10–30 cm/s, and model experiments have shown that they can be as large as 100 cm/s for extremely strong winds. Current speeds of about 60 cm/s have been observed in the Berg Strait. The winds lead currents by 12–20 hours, with the maximum lag observed in the deep western basin.

There were no measurements of bottom layer currents, but model results suggest that the bottom circulation sign is opposite to that of the surface currents, mainly because of barotropic geostrophic adjustment, with the "no motion" surface located at a depth of about 25 m in the western basin and 15 m in the central basin.

2.8   A BIT OF CHEMISTRY

2.8.1   Salt composition

The salt composition of the Aral Sea water is practically constant with respect to spatial coordinates, which is characteristic for water bodies of marine type (Blinov, 1956). The Aral Sea's ion composition is intermediate between the chloride–sodium-type ocean waters and hydrocarbonate–calcium-type continental waters. The content of principal ions in the Aral Sea water is displayed in Table 2.12.

The sulphate/chloride mass ratio SO_4/Cl is 0.90, which characterizes the Aral Sea as a sulphate-type water body strongly metamorphized by continental discharges.

2.8.2   Electrical conductivity

Because of the different chemical composition of the Aral Sea water and World Ocean water, their electric conductivities are different. The Aral Sea water conductivity is about 20% lower than that of the World Ocean water at the same temperature and salinity. This is mainly because of a higher relative content of SO_4^{-2} ions which are less mobile (Blinov, 1975). The following empirical formula has been used to calculate the conductivity κ (S/m) from salinity S (ppt) and temperature T (°C) (Sopach, 1958):

$$\kappa = (85.3 \cdot 10^{-3} + 2.487 \cdot 10^{-3}\,T + 0.160 \cdot 10^{-4}\,T^2)S$$
$$- (7.45 \cdot 10^{-4} + 0.94 \cdot 10^{-5}\,T + 0.590 \cdot 10^{-6}\,T^2)S^2 \tag{2.8}$$

This formula is applicable at an accuracy of 0.2% for 6.5 ppt $< S <$ 12.5 ppt and $0°C < T < 25°C$. Conductivity values obtained through the above equation have been tabulated (*Oceanological Tables*, 1964).

2.8.3   Oxygen and pH

The entire water body of the Aral Sea is well ventilated in all seasons. The absolute and relative (% of saturation) content of oxygen is usually maximum in the northern part of the Sea. Maximum content values are attained in the cold season, and the minimum in summer. The overall average content of oxygen in the surface layer is 6.27 ml/l, or 101% (Blinov, 1956). Most of the Sea is oversaturated with oxygen. The absolute minimum content (81%) was registered in June in the central part of the Sea at a depth of 25 m and in May in the bottom layer of the western trench. On the

Table 2.12. Content of major anions and cations in the Aral Sea water (45°95′N, 59°40′E, surface, June 1952).
From Blinov (1956).

Ion	Cl^-	SO_4^{-2}	HCO_3^-	Na^+	Mg^+	K^+	Ca^{+2}
Content (g/kg)	3.55	3.20	0.15	2.26	0.54	0.12	0.48

Table 2.13. Oxygen and pH in the deep western basin in October, 1950.

From Blinov (1956).

Depth (m)	T (°C)	S (ppt)	O_2 (ml/l)	O_2 (%)	pH
0	19.7	9.8	6.3	100	8.34
10	16.3	10.1	7.1	107	8.29
20	10.4	10.1	9.2	122	8.25
30	5.8	10.2	10.2	123	8.25
50	3.5	10.2	9.9	113	8.18
60	1.8	10.2	9.4	102	8.16

other hand, oversaturation up to 180% has been observed near the north-eastern shore. An example of vertical profiles of O_2 and pH are given in Table 2.13.

The value of pH of the Aral Sea water varies from 8.1–8.35. Spatially, it is distributed rather homogeneously, with small maxima characteristic for the areas adjacent to river mouths. The pH values tend to slightly increase with the depth in spring, but often decreases with the depth in summer and autumn (Table 2.13).

2.9   CONCLUSIONS

In the first-half of the past century, the Aral Sea was a large and rather special water body, combining both marine and limnic properties. Comparable with some seas by its spatial extent and dynamical characteristics, the Aral Sea also demonstrated peculiar physical behavior associated with elevated influence from continental fresh-water runoff, different salt composition implying different physical properties of the water, and rather specific bottom topography and hypsometry of the Sea. During this period, the hydrophysical budgets of the lake were equilibrated, which allowed for a quasi-stable state, being, however, subject to a considerable interannual variability. As in many lakes, because the salinity was low and not very variable either in time or space, many physical parameters were predominantly controlled by temperature rather than salinity (which is in sharp contrast with today's situation). On the other hand, salinity was an important agent in driving vertical and horizontal circulations of the Aral Sea. The water body was fairly homogeneous horizontally—although some differences between the basins (i.e., the western deep trench and the shallow eastern part, or the Large Sea and Small Sea), cannot be denied, these differences were relatively small, especially compared with those observed today; the vertical stratification was typically quite slim as well. The brackish lake was always well mixed and ventilated, factors which are perhaps the most generic features of the pre-desiccation Aral Sea, largely determining the character of its biological communities. Vertical mixing was partly sustained by thermal and haline convection which played a major role in the physical regime of the Sea. In winter, the convection processes were intimately connected with the ice formation. The near-bottom advection of denser water from the shallow eastern basin has been suspected to be partly responsible for the thermohaline structure in the deep layers. We shall see from what follows that this effect has greatly increased in the course of the desiccation.

3

Present-day desiccation

We now proceed to describing the present-day Aral Sea. Here, the present may be understood in a broader sense as the period after the early 1960s, when the desiccation processes had begun, but a special focus is made on the physical state of the lake in the last few years.

It must be said, once again, that this present state is largely unknown. As far as *in situ* measurements are concerned, observational data for the last decade are sparse. Doing any field research at the present Aral Sea is not an easy task, because the lake in its present state is rather remote and physically difficult to access and work in. The Sea has shrunk away from all roads, populated settlements, and infrastructure, including potable water sources. Navigation has ceased completely: at the time of writing, not a single vessel remained in permanent operation in the Large Aral Sea. Therefore, measurements can only be done from motor boats or similar portable platforms which must somehow be delivered to the site through hardly passable terrain, along with fuel, drinking water, and other necessary supplies. Although reduced, the lake still measures many tens and even hundreds of kilometers, and its surface is often rough, so collecting data at a reasonable spatial coverage from a small boat is time and fuel consuming and sometimes risky. Moreover, using conventional oceanographic instruments, such as Conductivity–Temperature–Depth (CTD) probes, in the Aral Sea water is not straightforward because of its differing salt composition, which implies different relations between conductivity, salinity, temperature, and density and thus poses serious additional difficulties for interpreting field measurements.

In what follows, we attempt to concisely summarize recent findings from remote sensing, numerical models, and direct measurements in the lake. The *in situ* hydrographic data are mainly based on the recent field campaigns of 2002, 2003, and 2004 (Zavialov et al., 2003b, c, 2004a; Friedrich and Oberhänsli, 2004).

3.1   LAKE LEVEL AFTER 1960

The Aral Sea level changes after 1950 are depicted in Figure 3.1. Regular data from direct measurements of the lake level were only available until 1991, however, there exist satellite altimetry reconstructions for 1993–2000 and direct geodesic determinations made in 2002–2004. Before the shallowing onset in 1961, the absolute elevation of the lake surface varied slightly around 53 m above the World Ocean level. Since 1961, the Aral Sea has been constantly shallowing at the rate of approximately 0.5 m/ year on average. The only three exceptional years when the level did not drop were 1969, 1970, and 2003.

To date (2004), the lake level has dropped by almost 23 m. The shallowing of the Aral Sea has led to dramatic changes to the shoreline, as shown in Figure 3.2 for 1978–2000. The present lake limit has moved many kilometers from its original position. A fortunate exception is the central western bank of the lake formed by the Ustyurt Plateau where the bottom slope is very steep and the retreat has been, therefore, relatively small. The largest horizontal withdrawal has occurred at the gently sloping eastern and southern extremities of the former bottom. By 1985, the Vozrozhdeniya, Komsomolskiiy, and Lazarev Islands have merged and formed a single large island between the deep western and shallow eastern basins of the Aral Sea connected through rather broad gaps south and north of the island. In 1989, the Berg Strait dried up and the southernmost portion of the lake (Small Aral Sea) detached from the main body (Large Aral Sea), forming an individual lake where the level has been relatively stable (with a variation of a few meters). The salinity in the Small Sea has varied from about 30 ppt immediately upon separation to about 20 ppt in 2002 (Friedrich and Oberhänsli, 2004). Following the separation, the Small

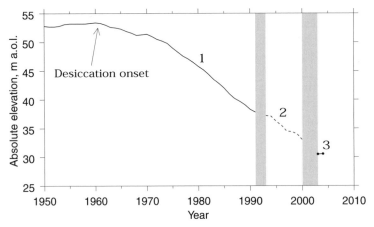

Figure 3.1. Long-term changes of the Aral Sea surface level (meters above ocean level): (1) historical data after Mikhailov et al. (2001); (2) TOPEX/Poseidon satellite altimetry reconstruction (http://www-aviso.cls.fr); (3) direct geodesic measurements in the surveys of 2002–2004 (Zavialov et al., 2003b, c, 2004a). Gray shading indicates gaps in the data.

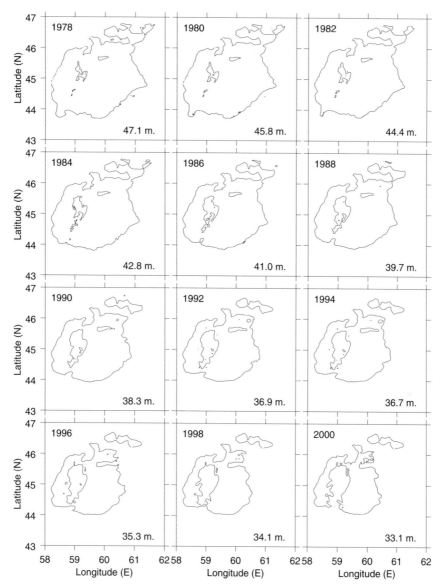

Figure 3.2. The shrinking lake: Aral Sea shoreline contours for 1978–2000 (the Small Sea after its separation from the Large Sea is shown schematically). The numbers at the bottom right of each plot are the absolute elevations of the water level above the mean World Ocean level (see Figure 3.1).

Sea level started to increase, while in the Large Sea shallowing continued. Small Sea water spilled at rates up to $3\,\text{km}^3/\text{year}$ (Aladin and Plotnikov, 1995a) into the Large Sea through the former Berg Strait, whose bottom had been dredged in the early 1980s. Fears have arisen that the erosion of the Berg Strait bed by the intense current could eventually affect the Syr-Darya mouth area and result in complete or partial diversion of the Syr-Darya inflow from the Small Sea into Large Aral. In August, 1992, a dike was constructed by the local Kazakh authorities to conserve the Small Sea and prevent its waters from leaking into the Large Sea. We note that damming of the Berg Strait had been previously suggested by a number of authors as a means of controlling the Small Sea level (e.g., Lvovich and Tsigelnaya, 1978; Chernenko, 1983). The dike has undergone a series of breaching and reconstruction episodes (e.g., Aladin and Plotnikov, 1995a). In 1997, the facility was replaced by a 20 km long, 26 m wide dam, which, however, has also breached in 1999. Since then, spillings from the Small Sea into the Large Sea have intermittently occurred at unknown rates. Plans for constructing a new, permanent dam are being elaborated (Micklin, 2004).

In 1996, the retreating shoreline of the Large Aral Sea had reached the former Barsakelmes Island in the eastern basin. Two years later, the large island between the two basins (sometimes still referred to as Vozrozhdeniya for convenience) had merged into the mainland in the south.

The mean depth of the Aral Sea had decreased from about 16 m in 1960 to 6 m in 2003. The lake surface area (66,100 km^2 in 1960) and volume (1,060 km^3 in 1960) have reduced to the respective values of 17,000 km^2 and about 100 km^3. Hence, the lake has lost over 90% of its water. Nonetheless, the Aral Sea still remains a notable water body whose maximum depth exceeds 43 m and horizontal extent is over 200 km.

Presently, the Large Aral Sea consists of two distinct basins connected by a single narrow and shallow channel (Figure 3.3, see color section). The width of the channel is about 3 km. The exact channel depth has been virtually unknown because the spatial resolution of available bathymetry is generally insufficient. According to some old bathymetric maps, the channel should not even be there now. It is sometimes hypothesized that the channel may have deepened because of the bed erosion in the course of the interbasin water exchanges which, as we show in Section 3.5, have been intense and played a major role in the immediate past. However, it has been believed that in the last few years the channel depth has not exceeded 1–2 m, which follows from the known bathymetry. The only available direct measurements taken in August 2004 partly disproved this notion and demonstrated that, at least at some locations, the channel was considerably deeper than expected, probably because of bed erosion (see Chapter 6 for details).

The western basin is a trench with a steep bottom slope at the western side where the maximum depth is about 43.5 m. The eastern basin is a relatively large but shallow hollow with a maximum depth of about 7 m.

3.2 WATER BUDGET COMPONENTS

Conventionally, the water budget equation is written as:

$$\frac{dV}{dt} = R - S(E - P) + G \qquad (3.1)$$

where V is the lake volume, t is time, R is the river runoff rate, S is the lake surface area, E is the evaporation rate, P is the precipitation rate, and G is the groundwater inflow rate. The bracketed terms in the right-hand side are sometimes jointly called the effective evaporation.

The equation can also be rewritten in the terms of the mean depth of the lake h. By definition, $h = V/S$, and the expression takes the form:

$$\frac{dh}{dt}\left(1 + \frac{h\,dS}{S\,dh}\right) = \frac{R}{S} - (E - P) + \frac{G}{S} \qquad (3.2)$$

3.2.1   River discharges

The runoffs of Amu-Darya and Syr-Darya into the Aral Sea for 1942–2001 are shown in Figure 3.4. The discharges are highly variable at the interannual scale, but do demonstrate a notable decrease by about one order of magnitude, from over $50\,km^3/year$ on average during the pre-desiccation period to only a few cubic kilometers per year in 1980s.

According to Bortnik (1996), the mean annual river inflow was $56.0\,km^3$ for 1911–1960, $43.4\,km^3$ for 1961–1970, $16.7\,km^3$ for 1971–1980, and $4.2\,km^3$ for 1981–1990. By the 1980s, the lake has gone far away from the former river mouths. Believed to be responsible for annual losses of up to $8\,km^3$ of the potential fluvial discharge (Létolle and Mainguet, 1996; Benduhn and Renard, 2004), water retention in the delta areas is among the factors limiting the freshwater inflow into Aral. The "delta retention" also makes it difficult to accurately assess how much of the river water actually reaches the lake, even if the water transport upstream of the former delta is known. However, estimates released by the Uzbekistan Hydrometeorological

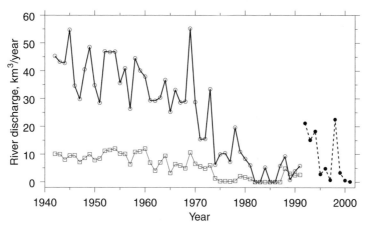

Figure 3.4. Annual discharges from Amu-Darya (bold curve, circles) and Syr-Darya (fine curve, squares) into the Aral Sea for 1942–1991, after Mikhailov et al. (2001), and estimated annual river discharges for 1992–2001 (dashed curve, bullets) obtained from the Uzbekistan Hydrometeorological Service.

Service (dashed curve in Figure 3.4) indicate that there was a moderate increase in the river runoffs up to about 9.0 km^3/year on average in the 1990s (see also Chub, 2000b).

3.2.2   Precipitation

The mean annual cycle of precipitation in the Aral Sea region for the desiccation period (1960–2001) is depicted in Figure 3.5. Also shown is the corresponding seasonal cycle of the ground air temperature. The meteorological data used in the plot are daily series from the Aralsk meteostation (46°47′N, 61°14′E) acquired from the Hydrometeorological Research Center of the Russian Federation. These data, as well as the corresponding NCAR/NCEP reanalysis data (e.g., Khan et al., 2004), reveal no significant long-term trends in the net precipitation over the desiccation period. The overall average for 1960–2001 is 110 mm/year, which coincides almost exactly with the average value for the pre-desiccation era (Bortnik and Chistyaeva, 1990). The shape of the curve, exhibiting two maxima (April and October) and two minima (February and September), is also essentially unchanged. The absolute contribution from precipitation in Aral's water budget, however, has been decreasing because of the reduction of the lake area. The total annual precipitation on the Aral

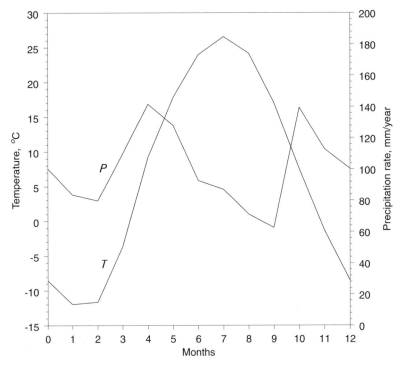

Figure 3.5. Mean annual cycle of precipitation (P) and air temperature (T) over the Aral region during the desiccation period.

Sea is estimated at $9.1\,km^3$ before 1960, $8.0\,km^3$ during 1961–1970, $6.3\,km^3$ during 1971–1980, and $5.5\,km^3$ during 1981–1990 (Bortnik, 1996; Benduhn and Renard, 2004). The estimated mean annual precipitation over the lake area at the time of writing is slightly below $2\,km^3$.

3.2.3 Evaporation

Only indirect estimates of the evaporation rates during the desiccation period are available, ranging from 900 mm/year (Björklund, 1999) or 970 mm/year (Bortnik, 1996) to over 1,200 mm/year (Benduhn and Renard, 2004) and even up to 1,700 mm/ year (Central Asian States, 2000). Some authors argue that there were no significant changes in the net evaporation rates during the desiccation (Bortnik and Chistyaeva, 1990; Bjorklund, 1999), but there are also strong indications that the rates may have been increasing. This is illustrated by Figure 3.6 showing annual effective evapora- tion against the Aral Sea volume (because E exceeds P by almost an order of magnitude, the effective evaporation is a good proxy for the evaporation itself). The values were obtained using Eq. (3.1) from the annual lake level and river discharges, while the groundwater exchanges were neglected. It can be seen that the effective evaporation calculated thereby, remained nearly constant at about 900 mm/year on average until the Aral Sea volume reduced to approximately $300\,km^3$ (around 1993), and then there was a marked increase to almost 1,600 mm/year. The overall regression over the entire desiccation period constitutes a 20% increase, incidentally confirming the result obtained by Small et al. (2001a) from a coupled regional climate–lake model. Possible causes and implications of this

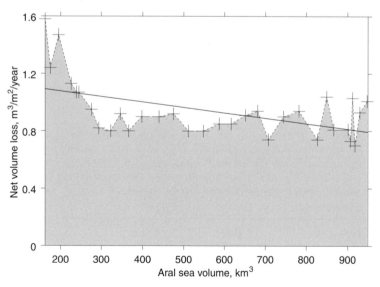

Figure 3.6. Effective evaporation (calculated as a residual term in Eq. (3.1) with G set to zero) during the desiccation period expressed as a function of the lake volume.

feedback will be addressed in Chapter 4. The estimated total annual evaporation has decreased from $66 \, km^3$/year in 1911–1960 to 15–$25 \, km^3$/year by the late 1990s.

3.2.4   Groundwater exchanges

At the time of writing, this water budget component is perhaps the most controversial. Very few quantitative estimates have been made, and these are indirect estimates largely diverging from essentially zero to tens of km^3/year. For example, Veselov et al. (2002) used a 3D dynamical model of the groundwater flow to the Aral Sea and arrived at 0.03–$0.06 \, km^3$/year. At the same time, Jarsjö and Destouni (2004) calculated G as a residual term in the water budget equation from the observed lake level changes and different evaporation scenarios for 1965–1996. For one of these scenarios, where the total evaporation rates were artificially maintained constant in the course of the lake desiccation, they obtained a groundwater inflow as large as $34 \, km^3$ in 1996. On the other hand, G varied between $-4 \, km^3$/year and $+4 \, km^3$/year for a more realistic but still somewhat artificial scenario where the evaporation per unit area rather than the total evaporation was fixed. In addition, Jarsjö and Destouni (2004) argued that the relative role of the groundwater discharges in Aral's water budget had increased compared with that in the pre-desiccation period, and the increase should be particularly pronounced in the western basin (see also Chapter 4). Finally, Benduhn and Renard (2004) used a water budget equation where the evaporation was parameterized through the modified Penman formula (Calder and Neal, 1984) rather than prescribed, and estimated G as $7.6 \, km^3$/year on average for 1981–1990. The representativity of this figure, however, depends on the adopted parameterizations and assumptions used and is therefore open to discussion.

Thus, it can be said that while the groundwater exchanges into the lake cannot be confidently quantified to date, they are likely to play a secondary but significant role in Aral's water balance.

The water budget components are summarized in Table 3.1.

Table 3.1. Summary of the water budget components (except groundwater discharge) after the desiccation onset in 1961. R is the annual fluvial discharge, P is the annual precipitation, and E is the annual evaporation. The first three lines are adapted from Benduhn and Renard (2004), the last line shows our estimates based on recent data.

Period	R (km^3)	P (km^3)	E (km^3)
1961–1970	43	8	65
1971–1980	17	6	55
1981–1990	4	5	39
1991–2001	9	1–4	15–25

3.3   SEA SURFACE TEMPERATURE, SALINITY, AND ICE

3.3.1   SST variability

At present, the main source of information about the variability of the SST and the
ice cover of the Aral Sea is satellite imagery. The temperature data used in this
section are the weekly Multi-Channel Sea Surface Temperature (MCSST) series
for 1981–2000 based on Advanced Very High Resolution Radiometer/National
Oceanic and Atmospheric Administration (AVHRR/NOAA) imagery (e.g.,
McClain, 1989; see also Ginzburg et al., 2003).

The weekly SST anomalies for 1981–2000 averaged over the entire Aral Sea area
are shown in Figure 3.7. These anomalies are highly variable and do not reveal any
significant overall trend. The data for the cold and warm seasons taken separately,
however, do exhibit trends of opposite signs: summer SSTs have been increasing at
the rate of $0.09 \pm 0.02°C$/year ($r^2 = 0.05$), while winter SSTs have been decreasing at
$0.14 \pm 0.02°C$/year ($r^2 = 0.11$), on average. The mean monthly SSTs (May, August,
and November) in different basins of the lake for the pre-desiccation period and late
1990s are presented in Table 3.2, where the data for 1994–2000 are satellite-derived
and those for the 1950s are historical data after Romashkin and Samoilenko (1953)
and Samoilenko (1955).

The SST changes which have occurred during the desiccation period are best
pronounced in spring and autumn—for instance, the present May SST is 4–5°C
higher than that in the 1950s, and the present November SST is 2–3°C below its
pre-desiccation value. Summer SSTs (August) have increased by over 2°C. Overall,
the annual SST range has increased from about 24°C to over 27°C and there is
also a notable phase shift for 3–5 weeks toward earlier maximum and minimum
(Figure 3.8).

Of course, the observed pattern of the desiccation-related SST changes is consistent with what should be expected for a shallowing water body whose heat storage
capacity is reduced because of both decreasing depth and increasing density stratification (to be discussed in Section 3.4). The observed changes fit reasonably well to
earlier predictions (e.g., Samoilenko, 1955a). The SST data also reveal growing
differences in the thermal regimes of the different parts of the Aral Sea. While
horizontal temperature gradients in the pre-desiccation state of the lake were
normally mild, at present, SST differences between the western and eastern basins
as large as over 5°C are not uncommon (Figure 3.9, see color section). An abrupt
increase of interbasin SST differences began in the early 1990s (Figure 3.10), coincident with the increase of the overall SST anomalies shown in Figure 3.7, and also
the increase of the effective evaporation rate (Figure 3.6). This suggests that the
beginning of the 1990s (the lake volume about $300\,km^3$, mean depth about 10 m)
was a critical point at which the shallowing process started to significantly affect the
depth of the thermally active upper layer, thus leading to consequent modulation of
the thermal regime of the Aral Sea. Before the 1990s, the SST variability was not
much different from that prior to the desiccation onset.

In addition to strong changes in the seasonal cycling of the SST, the
lake shallowing has apparently led to an increase in the diurnal SST range.

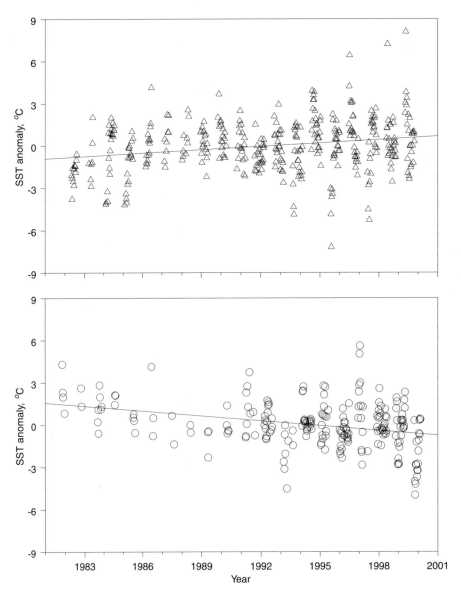

Figure 3.7. Satellite-derived weekly SST anomalies for 1981–2000. (*upper panel*) Anomalies for warm season (April–September). (*lower panel*) Anomalies for cold season (October–March). The straight lines are regression lines.

Day-to-night temperature variations with magnitudes above $3°C$ have been documented in recent surveys.

The diurnal changes in the thermal difference between the Sea and the surrounding lands are responsible for breezes, which have been investigated by hourly

Table 3.2. Monthly mean SST (°C) in different parts of the Aral Sea.
From Ginzburg et al. (2003).

Month	Period	Small Sea	Large Sea (western part)	Large Sea (eastern part)
May	1950s	11.9	11.3	12.3
	1994–2000	16.1	15.9	16.5
August	1950s	22.9	24.6	24.0
	1994–2000	25.4	26.1	26.3
November	1950s	6.0	11.0	9.1
	1994–2000	6.2	8.3	6.5

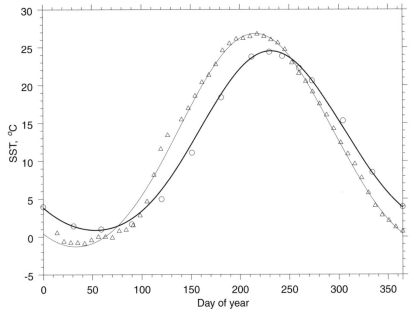

Figure 3.8. Mean annual cycle of Aral SST before 1981 (bold curve, circles) and after 1981 (fine curve, triangles). The circles are historical data after Bortnik and Chistyaeva (1990), the triangles are satellite-derived data after Ginzburg et al. (2003). The curves represent the sum of the annual and semi-annual Fourier terms best fitting the data.

launches of sounding balloons at the western shore of the western basin in surveys of 2003 and 2004. Typically, the onshore breeze appears at 2–4 m/s in the lower layer of the air (500–800 m) around 11:00 (local time) and then quickly expands vertically, reaching heights of 1.2–2.2 km by 14:00. The flow starts to decrease after 16:00. A weak offshore breeze can be observed during night-time and early morning hours at heights below 500 m (V. Khan, personal commun.).

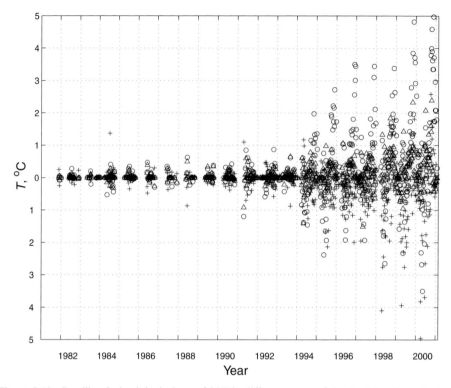

Figure 3.10. Satellite-derived deviations of SST in different parts of the Aral Sea from the SST averaged over the entire Sea: Small Sea (crosses), western Large Sea (triangles), eastern Large Sea (circles).

From Ginzburg et al. (2003).

3.3.2   Surface salinity

The salinity changes in the western and eastern basins of the Large Sea for 1960–2002, based on water samples collected from the surface, are reported by Mirabdullaev et al. (2004). These data combined with our data from the surveys of 2002–2004 are given in Table 3.3.

These data elucidate the general progress of the salinization whose rate has grown from 0.2 ppt/year at the initial stages of the desiccation to at least 5 ppt/year (western basin) and 40–50 ppt/year (eastern basin) in 2001–2002. Note that the abrupt increase of the salinization rates began in the mid-1990s. Simultaneously, the difference between the salinity values in the two basins which was essentially zero before 1996 also started to grow rapidly.

3.3.3   Ice regime

The ice regime changes during the desiccation period have been controlled by the changing thermal regime on the one hand and changing freezing point temperature

Table 3.3. Surface salinity (ppt) in the Large Aral Sea. The data marked by [†] are from Mirabdullaev et al. (2004), and those marked by [*] are from the author and his co-workers.

Year	Western basin	Eastern basin
1960	10[†]	10[†]
1970	12[†]	12[†]
1980	17[†]	17[†]
1990	32[†]	32[†]
1992	35[†]	35[†]
1995	42[†]	42[†]
1996	44[†]	44[†]
1997	49–51[†]	50–52[†]
1998	54[†]	58[†]
1999	56[†]	No data
2000	58–63[†]	No data
2001	63–68[†]	108–112[†]
2002	82[*]	155–160[†]
2003	86[*]	No data
2004	92[*]	100–110[*]

on the other. The latter factor, in turn, is determined by increasing salinity and changing salt composition. Water samples collected in a field survey of 2002 (Zavialov et al., 2003b, c) have been submitted to laboratory experiments aimed at determining the present-day freezing temperature and its dependence on salinity. The preliminary data from laboratory measurements at Shirshov Institute are shown in Figure 3.11. The left-hand panel refers to a water sample whose salinity was 88 ppt, characteristic for the present western Large Aral. The sample was cooled down and the sample temperature was recorded as a function of time. The acute minimum of the curve corresponds to over-cooled liquid, while the subsequent small maximum corresponds to ice formation. Therefore, in this case it can be seen that the freezing point of the sample was about $-3.8°C$. Similar values have been obtained previously in the experiments at the Marine Hydrophysical Institute, Ukraine (S. Stanichniy, 2004, pers. commun.).

The right-hand panel of Figure 3.11 shows the dependence of the freezing point on the salinity, as obtained from our experiments (Zavialov et al., 2004c). If extrapolated to higher salinities, the relation yields the freezing temperature of about $-7°C$ for the salinity 160 ppt reported for the eastern basin 2002 (Mirabdullaev et al., 2004).

Because of the abnormally low freezing temperature, at present, the ice cover of the lake is normally only partial, even in severe winters (Figure 3.12, see color section). Frequently, the Small Sea and the shallow northern portion of the Large Sea, including the channel connecting its western and eastern basins, are fully covered by ice, but the southern part of the lake is partly open, with pack ice extending only for 10–30 km from the shore in shallow waters and only a

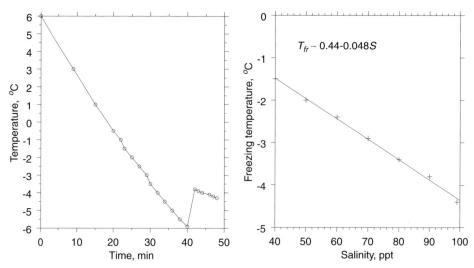

Figure 3.11. (*left panel*) Freezing curve for a sample at 88 ppt. (*right panel*) Freezing point as a function of salinity.

moderate amount of drifting ice in the central parts of the basins. A remarkable feature is the tongue of thick ice at the southernmost extremity of the eastern basin adjacent to the principal location of the Amu-Darya water inflow (see Figure 3.12, see color section). This structure is likely to have resulted from the encounter of fresh river waters and large masses of very cold—but still liquid—Aral Sea waters at temperatures below the freshwater freezing point, so the Amu-Darya runoff freezes because of cooling "from below" before it can be mixed into the hyperhaline environment. This phenomenon, which could be among the important mechanisms of ice formation in the present-day Aral, is frequently seen in recent satellite imagery (S. Stanichniy, personal commun.), but has not been quantitatively assessed to date.

A review of the present variability and recent changes of Aral's ice regime was given by Kouraev et al. (2004), based on satellite imagery. According to these data, the lake desiccation was accompanied by temporal shifts in the ice formation and melting at the annual scale, with the ice season starting about 15 days earlier than during the pre-desiccation period (mid-November to mid-January) and ending about one month earlier (late February), on average. The duration of the ice seasons for the period 1978–2002 varied in a broad range from 20–110 days depending on the severity of individual winters. This "irregular" variability largely masks possible trends related to desiccation and salinization. In the 1990s, there was a marked drop in the percentage of the lake area covered by ice from the maximum value 70% in 1992/1993 to the minimum below 10% in 1998/1999. These changes are likely to reflect larger scale climatic trends rather than the salinization effects, given that a similar ice cover reduction was observed for the neighboring Caspian Sea during this period (Kouraev et al., 2004).

3.4   THERMOHALINE STRUCTURE

As mentioned above, the vertical structure of the thermohaline fields in the present Aral Sea is poorly known because there have been very few, if any, hydrographic measurements in the bulk of the water column between 1992 and 2002. The lack of data is particularly severe for the eastern basin. The present thermohaline structure of the western basin is better known from five recent hydrographic surveys (Zavialov et al., 2003b, c; Friedrich and Oberhänsli, 2004; Zavialov et al., 2004a, b). The hydrographic transects referred to in the discussion below are depicted in Figure 3.13.

We begin with the thermohaline structure in the central, deep part of the western basin (Transect 1) documented in November, 2002. The corresponding vertical

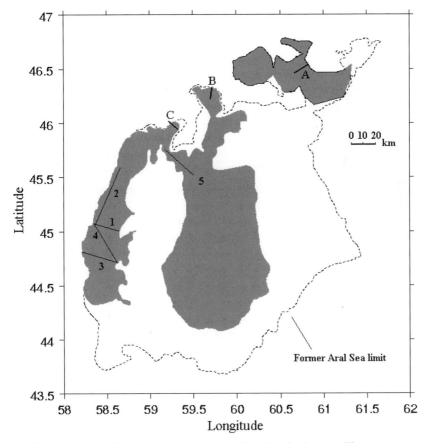

Figure 3.13. Locations of hydrographic transects referred to in the text. The transects marked by numbers were occupied by the author and his co-workers, and those indicated by letters were occupied by Friedrich and Oberhänsli (2004) and their co-workers.

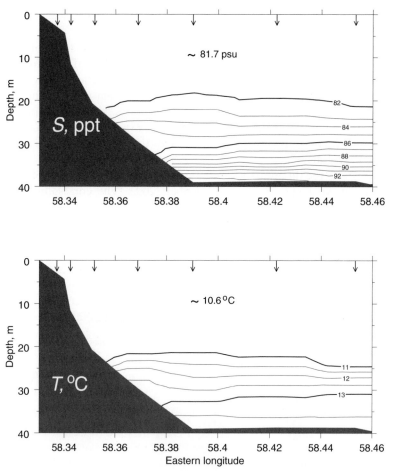

Figure 3.14. Vertical distributions of temperature T and salinity S at western portion of Transect 1 in November, 2002. Small arrows indicate locations of surface-to-bottom profiles used in the figure.

sections of temperature and salinity across the western basin from the west coast to the central axis of the trench are shown in Figure 3.14.

The measurement revealed the presence of very strong stratification in both salinity and temperature. We remember that no significant vertical stratification has been observed in the pre-desiccation period, and during the desiccation, until at least 1992 (except in the areas immediately adjacent to the Amu-Darya and Syr-Darya deltas). In the fall of 2002, the water column at all stations, where the depth was smaller than 20 m, was totally mixed, with salinity about 82 ppt and temperature about 10.6°C. The deeper stations exhibited a strong halocline where the salinity increased up to nearly 95 ppt at the bottom, accompanied by a strong temperature inversion. At the deepest stations, the bottom water was over 14°C (i.e., almost 4°C warmer than the surface water). A similar thermohaline structure was observed in August, 2002, by Friedrich and Oberhänsli (2004) in the Chernishov Bay in the northernmost part of the western basin (Transect C), where the salinity varied from 82 ppt at the surface to 110 ppt near the bottom (25 m), and the halocline was accompanied by a temperature inversion. At the same time, the water column was well mixed and relatively uniform in Tschebas Bay (Transect B) and Small Sea (Transect A) (Friedrich and Oberhänsli, 2004).

One can think of a kind of "three-layered" vertical structure of the western basin. The mixed layer extended down to approximately 20 m. Then there was a layer where the salinity changed with depth relatively slowly from 82 to about 86 ppt, while the temperature, in contrast, grew rapidly from the surface value to approximately 13°C. The thickness of this "intermediate" layer between roughly 20 and 30 m was about 10 m. Below, in the bottom layer, the salinity increase was very steep (up to 8 ppt per 10 m), while the temperature grew slowly. This three-layered structure indicated that at least three independent water types were present, which is also suggested by temperature–salinity (TS) analysis (see Section 3.5).

Picnometric measurements yielded the following density values: 1,056–1,057 kg/m^3 in the surface mixed layer, 1,060 kg/m^3 for the core of the intermediate layer as defined above, and 1,066–1,067 kg/m^3 for the bottom water. This means that there existed an extremely strong density stratification under the mixed layer. In the pycnocline, the density increased by approximately 0.5 kg/m^3/m, which corresponds to buoyancy frequency values as large as $N \sim 10^{-1}\,s^{-1}$.

Observations in the area were repeated almost exactly one year later, in the last 10 days of October, 2003. In the autumn of 2003, the measurements were done completely along the Transect 1 from the west bank to the "Vozrozhdeniya" coast in the east, and also at Transects 2–4 (Figure 3.13). The 2003 TS distributions at Transect 1, plotted in Figure 3.15, were much different from those observed a year earlier. First of all, the surface salinity value increased by about 4 ppt since the fall of 2002—even though there were no level drops during the wet year between the two surveys (Zavialov et al., 2004a). This local increase of the western basin salinity, therefore, can only be attributed to horizontal fluxes redistributing salt over the Sea area and, in particular, water exchanges between the two basins addressed in more detail in Section 3.5. Second, the strong temperature inversion seen in 2002 was no longer there (except in the near-bottom couple of meters at some stations, see below). Instead, the temperature generally decreased downwards, attaining values as low as 4.5°C in the deep layers. The structure adjacent to the western slope was also characterized by the highest salinity at this Transect (up to 93 ppt). The vertical distributions of T and S at the transversal Transect 3 (Figure 3.16) at the southern slope of the trench, where the lake is shallower, are similar to those in the upper portion of the water columns at Transect 1.

Particularly notable is the pattern of the distributions at the longitudinal segment along the western coast (Figure 3.17), synthesized from the data obtained at Transect 2 complemented with the data from the westernmost stations at Transects 1, 3, and 4. A sharp temperature front in the intermediate and deep layers is seen, separating a cold "lens" at the base of the southern slope from the

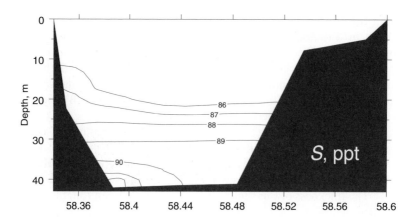

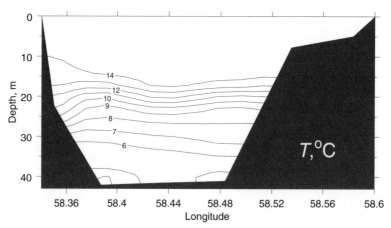

Figure 3.15. Vertical distributions of temperature T and salinity S at Transect 1 in October, 2003.

warm and relatively isothermal northern part of the basin. The cold core water in the southern part of the basin was not there a year earlier. The cold "lens" may have originated from either an eastern basin water intrusion in winter or, possibly, from winter cooling of waters in the shallow southernmost part of the basin accompanied by ice formation. The cold and saltier core waters having resulted from these processes may then have slipped downslope to their isopicnal level.

The haline structure shown in the upper panel of Figure 3.17 reveals a very high salinity (over 95 ppt) in the bottom layer. The pattern of the isolines in the bottom layer and the increase of both salinity and temperature northward is suggestive of a propagation of a near-bottom, warm and salty intrusion from north to south. A similar warm southward inversion could have been responsible for the thermohaline structure observed in the fall of 2002, given that the comparison of data shown in

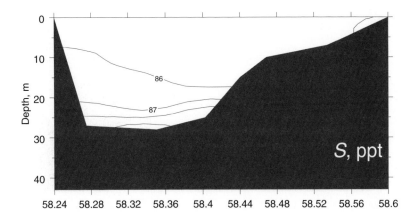

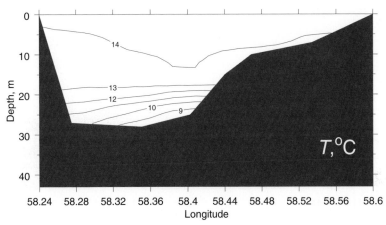

Figure 3.16. Vertical distributions of temperature T and salinity S at Transect 3 in October, 2003.

Figure 3.14 and the findings of Friedrich and Oberhänsli (2004) showed that both salinity and stratification increased to the north. We hypothesized that such intrusions originate from the shallow and salty eastern basin whose waters penetrate into the western trench through the connecting channel (Zavialov et al., 2003b, c). The role of such interbasin exchanges as virtually a major source maintaining the stratification in the western basin will be further elaborated in Section 3.5.

In some individual profiles of October, 2003, a hint of the presence of warmer and saltier water below the cold layer was evident in the lowest 3–4 m of the column (Figure 3.18). The characteristic "spikes" and stepwise structures in the profiles point to intense small-scale activity such as lateral advection events and, possibly, internal waves. Another important feature is the pronounced haline and thermal stratification observed in 2003 in the coastal shoals near the "Vozrozhdeniya" shore (Figure

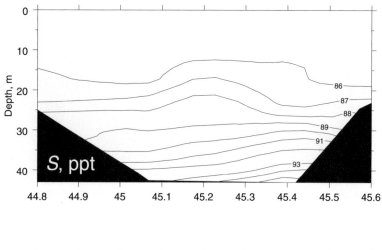

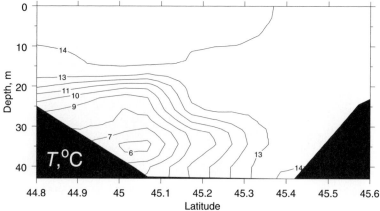

Figure 3.17. Vertical distributions of temperature T and salinity S along the longitudinal axis of western basin in October, 2003.

3.19), probably resulting from enhanced diurnal evaporation. Shallow banks of the western basin may therefore indeed be an important local source of saltier water which may then sink downslope.

The measurements of the hydrographic structure were repeated again in April, 2004, and then again in August 2004. However unlikely it may have seemed that the local convection could overturn such a strong salinity stratification observed in the fall of 2003 and bring surface water cooled down in winter to the bottom, the spring measurements at Transect 1 (April, 2004) have clearly demonstrated that the western Large Aral is not meromictic. The T and S profiles obtained in April are characteristic of a water column mixed by winter convection (Figure 3.20). This convection should be, at least partly, haline rather than purely thermal. Indeed, in oceanic conditions, it would be necessary to cool the surface water down by about $40°C$

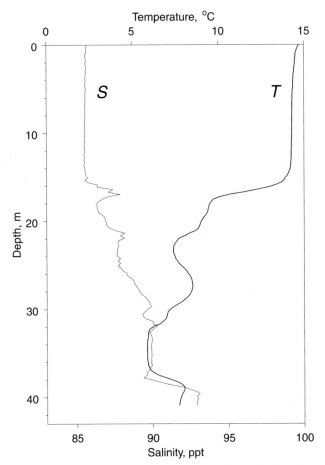

Figure 3.18. Typical vertical profiles of temperature T and salinity S taken on October 26, 2003, at 45°10′36″N, 58°23′24″E.

to override the initial surface-to-bottom salinity difference of over 10 ppt, which does not seem realistic. The thermal expansion and salinity contraction coefficients for the present Aral Sea water are unknown and the estimate may be different, but probably not by much. Hence, if the winter mixing does occur, it must be attributed, at least in part, to the haline convection, possibly connected with ice formation ("polar" type convection, as discussed in Section 2.4).

In August, 2004, measurements were taken at the standard Transect 1 and Transect 5, along the axis of the channel connecting the two basins of the Large Sea (Figure 3.13). In the deep western basin, the vertical pattern resembled neither the field observed in the fall of 2002 nor that seen in the fall of 2003. This time, the haline structure was inverse: the salinity *decreased* with depth from above 91 ppt in the mixed layer to about 88 ppt at the bottom (Figures 3.21 and 3.22). The water

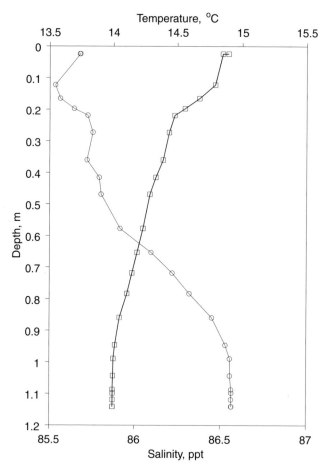

Figure 3.19. Highly stratified "Vozrozhdeniya" shoals: salinity (fine curve, circles) and temperature (bold curve, boxes) profiles at the point 44°43′42″N, 58°36′07″E, only a few tens of meters away from the eastern shore of the western basin; local depth 1.1 m. Reduced mixing and enhanced evaporation situation—warm and calm afternoon, October 26, 2003.

column remained stable because of the temperature drop of nearly 23°C between the surface and the bottom. We note that the surface salinity of the western basin increased by at least 4 ppt between April and August, and by nearly 9 ppt since 2002, although the lake level has not dropped since then (the geodesic leveling of November 2002 yielded an absolute surface level of 30.45 m, and that of August 2004, 30.75 m), so the local salinity increase should be attributed to fluxes of salt within the water body, in particular, the interbasin exchanges. As the measurements at Transect 5 have demonstrated, the salinity in the eastern basin (at least, in its northern part) was about 100 ppt, which indicates a marked drop compared with 155–160 ppt reported by Mirabdullaev et al. (2004) for 2002. This again suggests a

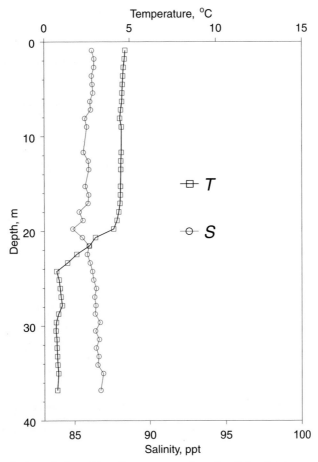

Figure 3.20. Typical vertical profiles of temperature T and salinity S taken on April 8, 2004, at $45°05'06''$N, $58°22'35''$E.

redistribution of salt excess from the eastern basin to the western trench. The surface salinity grew eastward along Transect 5 by about 4 ppt.

The elevated salinity observed in the mixed layer in August is due to enhanced evaporation in summer. Relatively low salinity in the bottom part of the column indicates that in the situation of August, 2004, no trace of a recent intrusion from the eastern basin was present. Unless such an intrusion occurs, the water column is doomed to overturn and become fully mixed in early autumn, as soon as the surface cooling results in a sufficient decrease of the temperature stratification.

Thus, the thermohaline structure of the Aral Sea which was fairly uniform in the pre-desiccation period (and largely remained so even over the course of the desiccation until the early 1990s) now exhibits a very strong vertical stratification and a rich horizontal variability. This variability is determined by: (i) lake–atmosphere interactions; and (ii) water, heat, and salt exchanges between the western and eastern

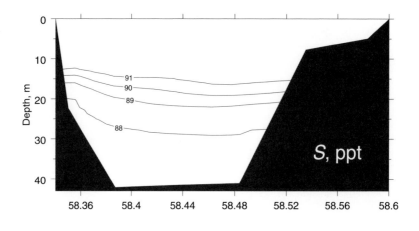

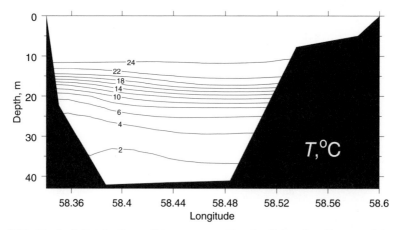

Figure 3.21. Vertical distributions of temperature T and salinity S at Transect 1 in August, 2004.

basins of the lake. Many of the related features manifested only in the subsurface layers are therefore "invisible" for remote sensing instrumentation.

It can be seen that the thermohaline fields and dynamical situations observed in the fall or late summer in three consecutive years are completely different from each other. In our opinion, this indicates that the physical state of the present Aral Sea is largely governed by "sporadic" variability, such as intermittent, wind-induced intrusions of the eastern basin water into the western trench, rather than "climatic" forcing, such as regular seasonal cycling. A major intrusion of salty eastern basin water in the bottom layer of the western basin creates strong stable stratification of the water column, which may persist for some period of time and lead to a great reduction in the vertical mixing. However, in the absence of the inflow of dense eastern basin waters, thermal and haline convection associated with enhanced

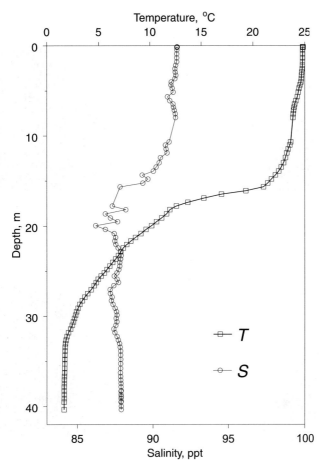

Figure 3.22. Typical vertical profiles of temperature T and salinity S taken on August 10, 2004, at $45°10'26''$N, $58°23'20''$E.

summer evaporation and winter cooling and ice formation are able to overturn the column and mix it. Consequently, the winter overturning may or may not occur in individual years, depending on the intensity of interbasin water exchanges on the one hand and the intensity of summer evaporation and winter cooling on the other.

3.5   WATER TYPES AND INTERBASIN EXCHANGES

We now develop a simple TS analysis to get a deeper insight into the thermohaline structure of the present Aral Sea. As known, this approach is widely used in classical oceanography to trace water types of different origins (e.g., Pickard and Emery, 1990), but has not been commonly applied to the Aral Sea.

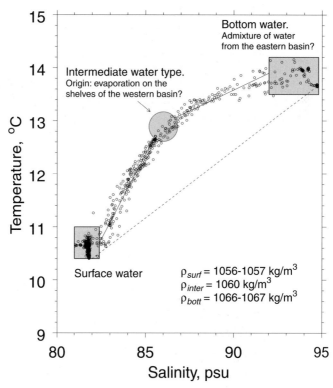

Figure 3.23. Temperature–salinity diagram for Transect 1, November, 2002. Gray shading indicates the basic water types, the straight lines are the regression lines for the respective portions of the diagram. The dashed line simply connects the diagram extremities to emphasize the diagram curvature. Also shown are the density values for the individual water types.

The data from a survey in the fall of 2002 plotted in TS coordinates are shown in Figure 3.23. It is not difficult to show that if two different water types (i.e., points on the TS plane) intermix, the point representing the resulting water type must lie on the straight line connecting the initial types, at the distances from them determined by the relative contents of the two types in the mixture. Therefore, inflection points of a TS diagram supposedly represent the "basic" water types, while the curve segments which can be approximated by straight lines symbolize mixing between these basic types. Further, if there are three basic types involved, any result of mixing between them generates a point lying inside the triangle known as the mixing triangle.

The shape of the TS diagram in Figure 3.23 suggests the presence of 3 distinct water types (i.e., the surface type, bottom type, and "intermediate" type) as schematized in the figure (the limits of their domains shown by the shaded boxes in the diagram were chosen subjectively based on examining all individual T and S profiles for different stations). Hence, the lower picnocline is essentially a mixture of the bottom and the intermediate water types, while the upper pycnocline represents

Colour plates

Figure I.1. View of the Aral Sea (western basin) from the Ustyurt Plateau in October, 2003.

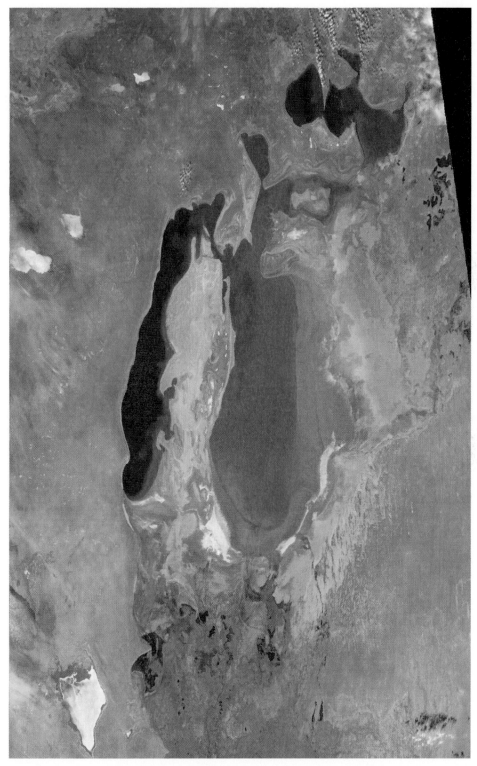

Figure 3.3. Present-day Aral Sea. MODIS AQUA (combination of Bands 1 and 2) satellite image taken on 23 October, 2003. True-color, 250-m resolution, 370×460 km.

Courtesy of S. Stanichniy, Marine Hydrophysical Institute, Ukraine.

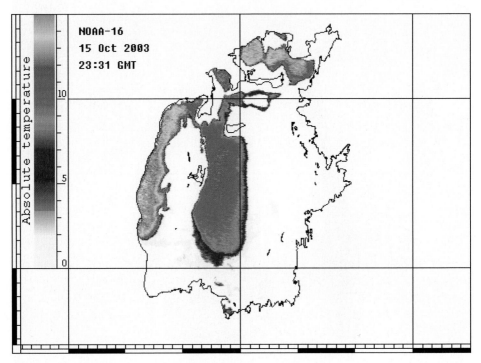

Figure 3.9. AVHRR Channel 4 (SST) image of the Aral Sea taken on 15 October, 2003. 1.1-km resolution, 630 × 440 km. Note the SST difference between the western and eastern basins.

Courtesy of S. Stanichniy, Marine Hydrophysical Institute, Ukraine.

Figure 3.12. Ice in the Aral Sea. MODIS AQUA (combination of Bands 1 and 2) satellite image taken on 22 February, 2003. True-color, 250-m resolution, 370 × 460 km.

Courtesy of S. Stanichniy, Marine Hydrophysical Institute, Ukraine.

Figure 3.31. "Swallow tails"—ancient concretions of precipitated gypsum found on the former bottom in October, 2003.

Gypsum deposits on the dry former bottom.

One of the motor boats used for recent field measurements.

the linear mixing between the surface water type from the upper layer and the intermediate water type.

The properties of the surface water type are determined mainly by lake–atmosphere interactions. As for the intermediate type, Zavialov et al. (2003b) hypothesized that it originates from the local evaporation in the western basin, especially on its shallow southern shelf where the summer evaporation is particularly intense. We further hypothesized that the bottom water type contains an admixture of waters originated from the eastern basin. Indeed, the bottom layer is warmer than the intermediate layer, which suggests that at least a part of the water constituting it attained its properties earlier, in the summer or early fall, when the SSTs in the shallow eastern basin are higher than anywhere in the western basin. The salinity in the bottom layer is much higher than in the intermediate layer, and it is unlikely that such a high salinity could have resulted from the evaporation in the western basin alone with no advection from the much saltier eastern basin. The waters from the eastern part of the sea were strongly suspected to contribute to the formation of the saltier deep waters of the western trench even before the desiccation onset (Simonov, 1962; Kosarev and Tsvetsinskiy, 1976).

The importance of the interbasin water and salt exchanges can also be illustrated by the following simple calculation. We consider the total salt content:

$$M = \int_0^h \rho S A \, dz$$

where h is the maximum depth, ρ is the water density, and $S(z)$ and $A(z)$ are the respective salinity and area of the horizontal cross section of the water body at depth z. In 1990, when the two separate basins had been already well developed, the total salt content in the western basin (defined as the part of the lake west of $59°E$) was about 3.8 billion tonnes, as follows from the data in Bortnik and Chistyaeva (1990). Calculating the total salt content in the same area in 2002 from the collected data and bottom topography, we obtain about 4.7 billion tonnes. Of course, this is only a crude estimate, because we assume that the salinity distribution is horizontally uniform. In fact, salinity at a fixed depth level may be different in different parts of the basin (cf. Friedrich and Oberhänsli, 2004), which could introduce an error to the calculated salt content. However, the calculation is likely to underestimate the total salt content rather than overestimate it, because, as far as the horizontal dimension at a fixed depth level is concerned, the survey area was probably the "freshest" portion of the basin, being located far away from shallow parts where most salinization occurs and from the channel connecting the basin with the hyperhaline eastern basin. This means that the salt content of the western basin has increased by at least 900 million tonnes between 1990 and 2002, or about 70 million tonnes/year on average. If the groundwater exchanges are neglected, this salt surplus (i.e., about 15% of the total mass of salts in the western basin) must have been advected from the eastern basin.

Two concurrent mechanisms which could be responsible for the penetration of the eastern basin water into the western trench are schematized in Figure 3.24. If the

Mechanism 1

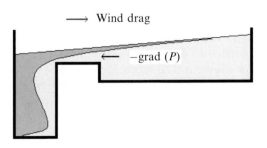

Mechanism 2

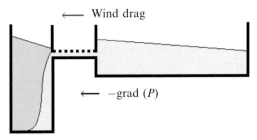

Figure 3.24. Conceptual schematic illustrating two possible mechanisms of eastern basin water intrusions into the western basin. Side look—the left vessel symbolizes the western basin filled with fresher water (dark gray), the right one is the shallow eastern basin with saltier water (light gray). (*upper panel*) The channel is sufficiently large so that the surface slope is established integrally. (*lower panel*) The channel is small so that the surface slope is established individually in each basin, and the two basins act as separate vessels, although connected through an imaginary fine "pipeline".

connecting channel is sufficiently large, then the eastward wind drag should affect the lake surface slope in such a manner that the dense eastern basin water is barotropic-ally pushed through the channel into the western basin where it sinks down to the bottom layers (upper panel). In the opposite conditions of westward wind drag, the vertical density pattern in and near the channel would be convectively unstable. On the other hand, in the limit where the channel is insignificant and the two basins "do not feel each other", the eastern basin water intrusion should be favored by easterly winds (lower panel in Figure 3.24). The system should eventually switch from the "large channel" scheme to the "small channel" one as the lake shallowing progresses. It is unclear which of the mechanisms is dominant at present, but model experiments suggest that the "large channel" mechanism is still active. In Figures 3.25 and 3.26, we present velocity fields obtained in our recent experiments with the well-known Princeton Ocean Model (POM) (e.g., Mellor, 1992) adapted to the Aral Sea (Rodin et al., 2005). The two fields correspond to the surface and the bottom currents, both obtained under westerly winds. As is clearly seen from the

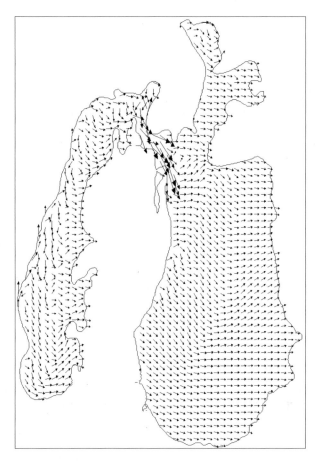

Figure 3.25. Surface velocity field obtained in a barotropic plus baroclinic mode experiment with the POM. The model used "real" stratification from Zavialov et al. (2003b, c) and was forced by constant and uniform westerly winds over 96 hours of the integration. Note the eastward flow in the surface portion of the connecting channel.

figures, while such winds drag surface water eastward through the connecting channel, they also do favor transport of the opposite sign in the bottom portion of the channel. The model therefore was able to reproduce the "large channel" mechanism as described above.

We now try to quantify the role of the interbasin water exchanges. We assume that the bottom water type (S_B, T_B) observed in the western basin in November, 2002, is actually a mixture of the intermediate western basin water (S_W, T_W) and the eastern basin water (S_E, T_E). If the water density and specific heat are assumed constant (which, in this case, is a good approximation), the heat and salt conservation equations read:

$$VT_E + (1 - V)T_W = T_B \qquad VS_E + (1 - V)S_W = S_B \qquad (3.3)$$

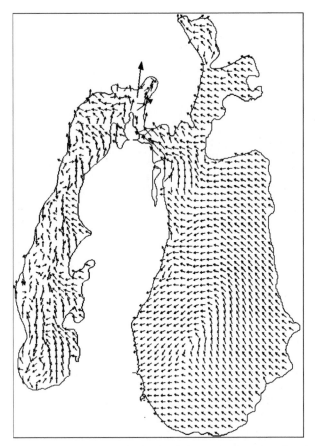

Figure 3.26. Near-bottom velocity field obtained in a barotropic plus baroclinic mode experiment with the POM. The model used "real" stratification from Zavialov et al. (2003b, c) and was forced by constant and uniform westerly winds over 96 hours of the integration. Note the westward flow in the bottom portion of the connecting channel.

or

$$V = \frac{T_B - T_W}{T_E - T_W} \qquad S_E = \frac{S_B - S_W(1 - V)}{V} \qquad (3.4)$$

where V is the volume fraction of the eastern basin water in the western basin bottom water. Based on the observations, we set $T_B = 13.9°C$, $T_W = 12.8°C$, $S_B = 93.5\,ppt$, and $S_W = 85.5\,ppt$. Hence, if we can estimate the initial temperature T_E of the eastern basin water intrusion from some independent considerations, we then obtain the relative content V of the eastern basin water in the bottom layer and, incidentally, estimate the eastern basin salinity S_E. Assuming that the intrusion was forced through mechanism 1 explained above and considering that, according to the wind record for 2002 the last major event of favorable winds before the observations

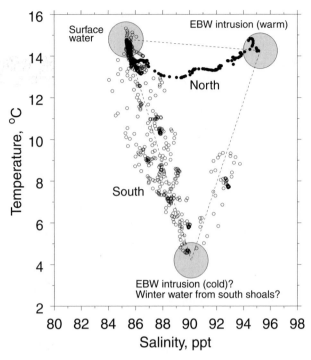

Figure 3.27. Temperature–salinity diagrams for the southern part (circles) and northern part (bullets) of the western basin in October, 2003. Three basic water types are identified (large gray circles). EBW stands for Eastern Basin Water.

was in August, we hypothetically attribute the intrusion to August, 2002. Infrared satellite imagery revealed that at that time, the eastern basin temperature varied between 22°C and 25°C, so we explore this range of T_E. The corresponding interval for V is 0.09–0.11, meaning that about 10% of water in the western basin bottom layer originated from a recent intrusion of the eastern basin water. The estimated total mass of salt advected into the western basin during this intrusion is between 12 and 16 million tonnes. The estimated interval for S_E is from 158 ppt to 179 ppt. We note that the latter estimate for the eastern basin salinity agrees quite well with the value suggested by Mirabdullaev et al. (2004) (see Table 3.3).

The TS diagrams for October, 2003, are shown in Figure 3.27. Three basic water types forming a "mixing triangle" can again be identified. The thermohaline properties of the new warm intrusion are close to those observed in 2002. According to the wind record, the 2003 intrusion is most likely to have occurred in June. The origin of the cold water type is arguable and could be associated with either earlier winter intrusion from the eastern basin or, possibly, winter convection or downslope penetration of very cold winter water from the shallow southern part of the western basin.

In summary of this section, it can be said that the TS analysis points to an extreme importance of the water and salt exchanges between the two basins.

Intrusions of very salty and dense eastern basin water through the connecting channel are likely to have played a central role in establishing the thermohaline regime and vertical stratification in the western trench. If the lake desiccation continues at 1990s rates, the channel risks drying up completely in the next few years and then the interbasin exchanges will cease. The channel closure should have a major impact on thermohaline regimes in both basins, which in turn may significantly alter further desiccation progress. Some of the related issues are addressed in Chapter 4.

3.6   CIRCULATION

The drastic changes in the shoreline shape, bathymetric characteristics of the lake, and its thermohaline structure must have resulted in significant alteration of Aral's circulation patterns. The present circulation, however, is largely unknown—the only direct measurements of currents reported for the desiccation period are confined to a single point and a short period of time (Zavialov et al., 2004a). Beyond this isolated observation, we must rely on indirect estimates from satellite information and numerical models which are not numerous.

Using a barotropic 3D numerical model, Barth (2000) and Sirjacobs et al. (2004) investigated the circulation pattern in 1981–1985, a very low river inflow period. They argued that the annual mean basin-scale circulation pattern at that stage of the desiccation remained anticyclonic in both basins of the Large Sea as well as in the Small Sea. However, they noticed an intensification of seasonal changes in the circulation, and in winter, the main anticyclonic gyre was least developed, and moreover cyclonic circulations appeared in the deep western basin and the north-eastern part of the eastern basin. This notion, as well as some satellite-derived velocity snapshots such as the cyclonic field in Figure 3.28, led some authors to hypothesize that the Large Sea's circulation pattern may have switched from anti-cyclonic to cyclonic, at least in autumn and winter, in the course of lake shallowing and salinization (e.g., Zavialov et al., 2003b). But this idea does not seem to have been supported by our recent modeling studies which used the well-known POM (e.g., Mellor, 1992) adapted to the Aral Sea conditions (Rodin et al., 2005). Barotropic experiments as well as those with the real stratification showed that the mean circulation at the surface forced by climatic winds remained anticyclonic in either basin (Figure 3.29).

A series of recent direct velocity measurements in the Aral Sea is presented in Figure 3.30. The data were collected in October, 2003, at two consecutively occupied mooring stations equipped by mechanical current meters (Zavialov et al., 2004a). The two stations were located only 300 m apart from each other at the slope near the western coast of the western basin (45°06′N, 58°21′E). The instruments were deployed at depths 7 m (Station 1) and 15 m (Station 2). The series suggests the anticyclonic character of the large-scale circulation, the predominant velocity at the mooring site being directed to the north-east, even against the north-easterly wind. The subsurface currents in the stratified environment of the Aral Sea exhibit

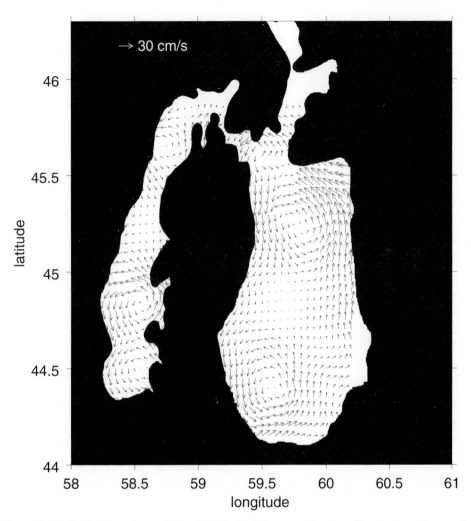

Figure 3.28. Cyclonic surface velocity field derived from a pair of satellite images taken on 9 and 10 November, 2002, by the Maximum Cross-Correlation (MCC) technique (e.g., Zavialov et al., 2002b).

little correlation with the wind stress at the surface. Figure 3.30 also gives an idea of the velocity values characteristic for the Sea in this recent observation, with the mean velocity about 3 cm/s and the maximum velocity 12 cm/s.

3.7　A BIT OF CHEMISTRY

3.7.1　Salt composition

The salt composition of the Aral Sea has changed during the desiccation. The principal process responsible for these changes is chemical precipitation of some

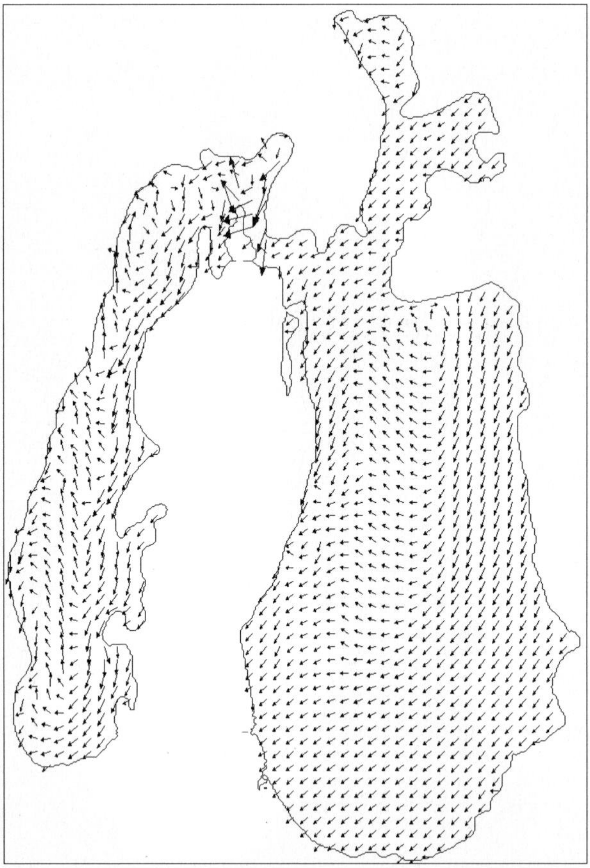

Figure 3.29. Anticyclonic surface velocity field obtained in a barotropic plus baroclinic mode experiment with the POM. The model used "real" stratification from Zavialov et al. (2003) and was forced by constant and uniform north-north-easterly winds over 96 hours of integration.

compounds accompanying the salinization. The salinity increase leads to a decrease of carbon dioxide solubility in the water (Bortnik and Chistyaeva, 1990) and therefore the dioxide equilibrium shifts towards the passage of HCO_3^- to CO_3^{2-}, which results in precipitation of the excessive CO_3^{2-} in the form of the calcium carbonate $CaCO_3$:

$$Ca(HCO_3)_2 \rightarrow CaCO_3 \downarrow + CO_2 + H_2O \tag{3.5}$$

For higher salinities, magnesium carbonate $MgCO_3$ is also precipitated in a similar reaction. According to Bortnik and Chistyaeva (1990), this process had been responsible for the precipitation of about 100 million tonnes of carbonates between 1960 and 1989. At the initial stages of the salinization, accumulation of sulphates in the form of $CaSO_4$ occurred. The subsequent salinization has led to

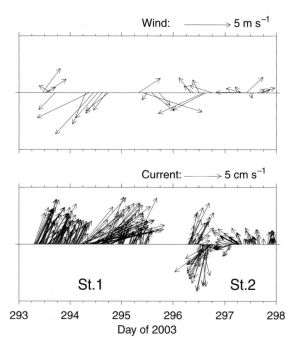

Figure 3.30. (*lower panel*) Velocity vectors ("stick diagram") measured at two stations in October, 2003. (*upper panel*) Corresponding wind record. The current velocity and wind speed scales are given at the top of the plots.

oversaturation of Aral Sea waters with calcium sulphate and precipitation of gypsum $CaSO_4 \cdot 2H_2O$:

$$Ca(HCO_3)_2 + MgSO_4 \Leftrightarrow CaSO_4 + Mg(HCO_3)_2 \tag{3.6}$$

$$CaSO_4 + 2H_2O \rightarrow CaSO_4 \cdot 2H_2O \downarrow \tag{3.7}$$

This process must have started when the salinity exceeded 26–28 ppt (Bortnik and Chistyaeva, 1990). Deposits of gypsum precipitated during the ancient regressions of the Sea are commonly seen on the former bottom, sometimes having the form of individual concretions called "swallow tails" for their specific shape (Figure 3.31, see color section).

Other processes involved include the precipitation of mirabilite $Na_2SO_4 \cdot 10H_2O$ starting at salinity values around 150 ppt (e.g., Rubanov et al., 1987), glauberite $CaSO_4 \cdot Na_2SO_4$, and epsomite $MgSO_4 \cdot 7H_2O$ at higher salinities.

The present ion content in the western basin water obtained from water samples collected in the survey of November 2002, is summarized in Table 3.4. During the survey, pH at the surface varied between 8.0 and 8.5, which constitutes an increase compared with the pre-desiccation period.

The temporal changes in the concentrations of the principal ions are illustrated in Figure 3.32, where the historical data for 1952 and 1985 are from Bortnik and Chistyaeva (1990) (see also Drumeva and Tsitsarin, 1984).

Table 3.4. Present content of major anions and cations in the western Aral Sea water. The water samples analyzed were taken in the survey of November 2002 (Zavialov et al., 2003b, c). The analyses were done at the Abdoullaev Institute of Geology and Geophysics, Uzbekistan.

Ion	Cl^-	SO_4^{2-}	HCO_3^-	Na^+	Mg^{2+}	K^+	Ca^{2+}
Content (g/l)	35.0	24.1	0.6	21.9	5.0	0.9	0.6

Hydrochemical properties of the present Aral Sea water were investigated and first reported by Friedrich and Oberhänsli (2004) based on samples taken in the Small Sea (Transect A) and the northernmost tip of the Large Sea (Transects B and C) in summer 2002. They reported a rather uniform chemical composition of the water column in the Large Sea, with the SO_4/Cl mass ratio varying between 0.77 at the surface and 0.75 in the bottom layer, while in the Small Sea, the ratio was spatially uniform at 1.02. The data presented in Table 3.4 are generally consistent with the data for the Large Sea published by Friedrich and Oberhänsli (2004). As mentioned in Chapter 2, the SO_4/Cl ratio is one of the principal chemical state indicators essentially determining the chemical type of a water body. The pre-desiccation (1952) value of the mass ratio for the Aral Sea water was 0.90 (Blinov, 1956). The observed decrease of the ratio quantifying the metamorphization processes mentioned above had been predicted (Bortnik and Chistyaeva, 1990), but the values observed in 2002 and 2003 in the Large Sea are about 20% higher than those previously expected for today's salinity. In contrast, in the Small Sea where the salinity was 18–20 ppt, the SO_4/Cl ratio actually increased compared with the pre-desiccation value. Elevated values of the ratio are characteristic for continental discharges, so this increase may point towards significant metamorphization of the Small Sea by residual Syr-Darya runoffs.

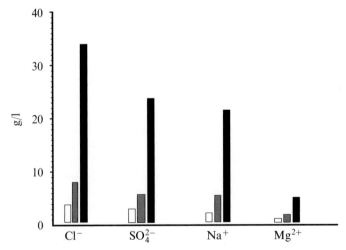

Figure 3.32. Content of principal ions in the Large Aral Sea water in 1952 (white), 1985 (grey), and 2002 (black).

3.7.2   Anoxic zone and hydrogen sulphide

One of the most spectacular findings from the recent field work in the Aral Sea was the discovery of a huge anoxic and hydrogen sulphide contaminated water body below the mixed layer in the western basin of the Large Sea observed in 2002 and 2003 (Zavialov et al., 2003b, c; Friedrich and Oberhänsli, 2004). Of course, this anoxic zone should be intimately connected with the strong salinity and density stratification which greatly reduces vertical mixing. In the pre-desiccation Aral Sea, the entire water body was ventilated and there was no sign of H_2S in the water, although traces of the hydrogen sulphide in the bottom silts have been long known (e.g., Bogomolets, 1903; Blinov, 1956). Apparently, H_2S first appeared only recently—no earlier observers reported it.

Typical vertical profiles of H_2S and O_2 concentrations are shown in Figure 3.33. Oxygen vanishes at a depth of about 20 m, where hydrogen sulphide immediately appears. Its concentration increases downwards and attains maximum values of over 80 mg/l, which is, for example, one order of magnitude higher than H_2S concentrations characteristic for the Black Sea (e.g., Volkov et al., 2002).

The approximate horizontal extent of the H_2S contaminated zone in November 2002, estimated under the assumption that its upper limit lies at a depth of 20 m is

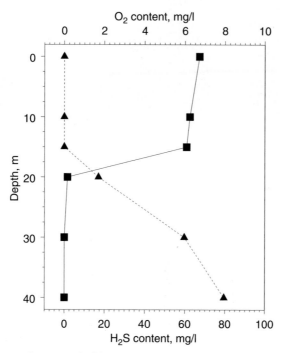

Figure 3.33. Content of oxygen (solid curve, boxes) and hydrogen sulphide (dashed curve, triangles) in the Aral Sea water in October, 2003. The profiles were acquired at 45°04′04″N, 58°23′14″E.

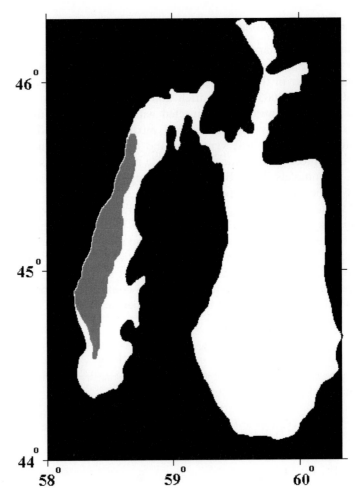

Figure 3.34. Estimated spatial extent of the H$_2$S contamination zone in 2003 (gray).

sketched in Figure 3.34. In fact, the actual extent is probably larger in the north, given that the mixed layer depth decreased northward and the presence of H$_2$S in the bottom layers of the Chernishov Bay at the northern extremity of the western basin (Transect C) was reported by Friedrich and Oberhänsli (2004) for the same period. The total content of hydrogen sulphide at that time is estimated as 500,000 tonnes.

Surprisingly, the observations made in spring of 2004 showed no traces of H$_2$S throughout the water column which was relatively well mixed. In August 2004, H$_2$S was present, but only in the bottom few meters of the column and in a very low concentration (as subjectively estimated from smell, quantitative figures are not available). Therefore, the H$_2$S contamination in the western Large Aral is a seasonal or intermittent rather than permanent feature. Does Aral's anoxia occur annually on a seasonal basis followed by winter mixing, or has the lake undergone a "meromictic" period during several years prior to 2003, leading to a long-term accumulation of H_2S?

The biochemical origins allowing for such a rapid build up of the H_2S zone in the Aral Sea is an important subject for future research.

3.8 DENSITY AND ELECTRICAL CONDUCTIVITY

As discussed in Chapter 2, the electrical conductivity and density of the Aral Sea water are different from those of the World Ocean water at the same temperature and salinity because of a different salt composition. This was the case prior to the desiccation onset, and therefore the corresponding empirical formulas specific for the Aral Sea were used at the time (Blinov, 1956; Sopach, 1958). These formulas, however, were obtained only for a rather narrow salinity range typical for the "old" Aral Sea, and, of course, were not intended for the present Sea whose salinity has increased by almost one order of magnitude and whose salt composition has also changed considerably. No comprehensive formulas or tables adjusted for the present conditions are known. In Figures 3.35 and 3.36, we present some results of conductivity and density measurements made in the laboratory (Shirshov Institute of Oceanology) on the water samples collected in the western Large Aral in 2002 and 2003. In these experiments, the initial salinity was determined chemically using the "dry residual" method through calcination as described by Blinov (1956). Lower salinity samples were then prepared by diluting the original water. Also plotted are the lines obtained by formally extrapolating the corresponding "old Aral" equations. It might be of interest to compare these extrapolations with the measurements, given that these equations are sometimes used for the present Aral Sea, simply due to lack of an update.

Linear regressions of the laboratory data yielded the following approximate relations for the density σ_{20} (kg/m^3) and conductivity C_{17} (S/m) at the respective temperatures 20°C and 17°C:

$$\sigma_{20} = -3.8 + 0.75S \qquad 60 \le S \le 93 \tag{3.8}$$

$$C_{17} = 0.44 + 0.101S \qquad 20 \le S \le 93 \tag{3.9}$$

where S is the salinity (ppt). Both regressions fit the available data at $R^2 > 0.99$ with the r.m.s. errors of 0.2 S/m for the conductivity and 1.0 kg/m^3 for the density. Obtaining more accurate relations, including the dependences on the temperature, is subject to future research.

3.9 OPTICAL PROPERTIES OF WATER

The optical transparency of the western basin water as measured by Secchi disk in the surveys of 2002–2004 varied between 3 and 7 m, constituting a notable decrease compared with the pre-desiccation values typical for the location (Chapter 2). At

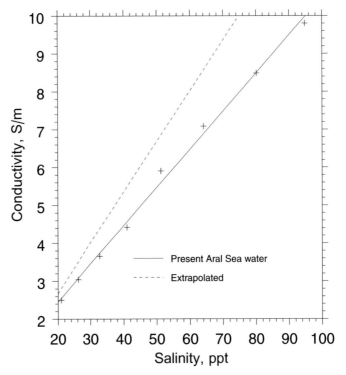

Figure 3.35. Electric conductivity (S/m) of the Aral Sea water as a function of salinity, as obtained from laboratory measurements (crosses, the solid line is the linear regression), and formally calculated from the "pre-desiccation" empirical formula (Sopach, 1958) (dashed line). The temperature is constant at 17°C.

present, the transparency in the basin tends to decrease from west to east, and minimum values are observed near the eastern shore. The dominant colors of water are the greenish-blue tones (i.e., color values 6–7 in the color scale used by Bortnik and Chistyaeva (1990)), which are different from the typical pre-desiccation values of 4–6.

In the channel connecting the two basins of Large Aral and in the northern part of the eastern basin (Transect 5) in August, 2004, the Secchi depth was 1–1.5 m and the water color was yellowish-green and greenish-yellow (i.e., 11–15).

3.10 CONSEQUENCES OF DESICCATION

The Aral Sea desiccation has triggered a variety of environmental consequences at the regional scale. First of all, the biological communities of the lake and surrounding area have suffered dramatic degradation. Presently, the biodiversity of algae species in the plankton is below 50% of that documented before the desiccation

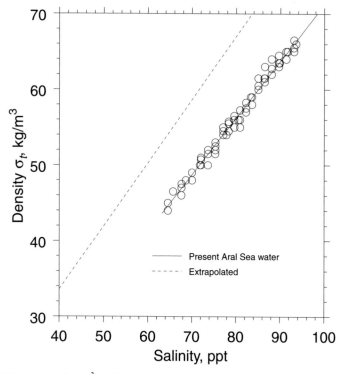

Figure 3.36. Density σ_t (kg/m^3) of the Aral Sea water as a function of salinity, as obtained from laboratory measurements (circles, the solid line is the linear regression), and formally extrapolated from the "pre-desiccation" empirical formula (Blinov, 1956) (dashed line). The temperature is constant at 20°C.

onset, the total number of phytoplankton taxa has decreased from 306 in 1974 to 159 in 2002, the number of zooplankton species has reduced from 42 in 1971 to 4 in 2002, and that of zoobenthos from 67 in 1970 to only 3 in 2002 (Mirabdullaev et al., 2004). With once numerous fish species, only 2 (*Atherina boyeri caspia* and flounder *Platichtys flescus luscus*) were recorded in 2002 (Mirabdullaev et al., 2004, Zavialov et al., 2003c). Since 2000, the halophyllic brine shrimp *Artemia salina* is the dominant zooplankton species accounting for 99% of the biomass. It is expected to become the only animal in Aral in the immediate future. However, the biological consequences of the Aral Sea desiccation which have been addressed in detail elsewhere (e.g., Mirabdullaev et al., 2004; Aladin et al., 2004) are beyond the scope of this book.

Another important group of consequences are the regional climate impacts. The available estimates of the Aral Sea basin climate change differ significantly. According to Muminov and Inagatova (1995), the monthly mean surface air temperatures in the region have increased by up to 6°C in summers while Chub (2000a) argued that the lake desiccation may be responsible for an air temperature anomaly of 1–1.5°C. The Aral Sea desiccation impacts are uneasy to separate from larger scale climate change. Indeed, regionally coherent air temperature trends not related to the Aral Sea shallowing have been observed all over Central Asia between 1960 and 1997. Statistical approaches aimed at isolating the desiccation-related climate changes were developed by Small et al. (2001b) and Khan et al. (2004). Such methods have led to the conclusion that the changes in the surface mean air temperature specific to the Aral Sea region were as large as 2–6°C from the period after 1960. The magnitude of the changes was largest along the former shoreline and decreased with the distance from the lake (Small et al., 2001b). As may have been expected, the changes were most pronounced south-west of the Aral Sea, down the predominant north-easterly winds. Warming was detected during summer and spring and cooling during winter and autumn, which suggests a weakening moderating effect from the shallowing lake. The spatial extent of the temperature changes is about 200 km in the downwind direction. Finally, an increase of the diurnal temperature range of 2–3°C was observed near the Aral Sea in all months (Small et al., 2001b). A similar pattern of air temperature changes south-west of the Aral Sea was independently obtained by Khan et al. (2004) from the National Center for Atmospheric Research/National Center for Environmental Prediction (NCAR/NCEP) reanalysis data not only for the land surface but also for higher levels in the atmosphere. They concluded that in the vertical direction, the desiccation-related trends in the air temperature can be traced up to 700 mb (i.e., an altitude of about 3 km). The impacts of the Aral Sea desiccation on other parameters of the regional climate such as atmospheric precipitation, air humidity, and wind regime remain poorly explored (see also Section 4.1.1).

The desiccation of the Aral Sea has caused a number of other negative effects which must be briefly mentioned here. The retreat of the shoreline has led to the formation of extensive salt marshes. As a consequence, salt and dust are introduced into the atmosphere by wind erosion and transported 200–500 km and sometimes even farther (Micklin, 2004). Some estimates indicate that every km^2 of the former bottom produces about 8,000 tonnes of salty dust which then falls down on the adjacent lands, thus leading to salinization and degradation of soils (Kiselev, 2004). Intense dust and salt storms in the Aral Sea region were first observed by means of satellite imagery in the mid-1970s and their incidence has been increasing since then. The estimated net aeolian transport of salt from the dried bottom was around 40 million tonnes/year in the 1980s (Rubanov and Bogdanova, 1987).

3.11   CONCLUSIONS

A summary of the changes in the physical oceanographic state of the Aral Sea after 1960 through to 2003 is given in Table 3.5.

In our view, the entire desiccation period (1960–present-day) can be divided into two intervals. The first, encompassing about 3 decades (1960–late 1980s), could be called the "early desiccation period". It was characterized by gradually

Table 3.5. Summary of the changes in the physical oceanographic state of the Aral Sea, from 1960 to 2003.

Parameter	Pre-desiccation	2003
Volume (km^3)	1,060	95 (Large Sea) 23 (Small Sea)
Area (km^2)	66,000	16,000 (Large Sea) 3,000 (Small Sea)
Mean depth (m)	16	6
Salinity (ppt)	9–10	~ 20 (Small Sea) ~ 90 (Large Sea, western basin) 100–150 (Large Sea, eastern basin)
Density (kg/m^3)	1,005–1,009	1,056–1,067 (Large Sea, western basin)
Freezing point (°C)	−0.6	−4.3 to − 3.6 (Large Aral, western basin)
Annual SST range (°C)	~ 24	Over 27, phase shifted for 3–5 weeks
Vertical density stratification (kg/m^4)	Slim (<0.1) or none, except near the river mouths	Strong, up to 1
Horizontal gradients	Small	Interbasin differences up to 7°C in temperature and 70 ppt in salinity
Evaporation rates	Nearly constant	Elevated by 10–20%
Circulation	Anticyclonic	Anticyclonic(?), with cyclonic vorticity manifestations in autumn and winter
Chemical composition	High content of SO_4^{2-}	SO_4/Cl mass ratio decreased by 20% in Large Sea
Ventilation	Fully ventilated	Bottom layer anoxic, H_2S present in summer–autumn at concentrations up to 80 mg/l

accumulating, relatively slow and uniform, and fairly linear changes in the physical properties of the water body with time. At that stage, the thermal regime of the active upper layer of the lake, as well as the vertical stratification pattern below it, were altered only slightly and, as far as the physical oceanographic state is concerned, the Aral Sea essentially "did not feel" the ongoing desiccation. The annual evaporation rates were nearly constant. Rapid transition to a new regime occurred in the early 1990s. The subsequent "severe desiccation" period started when the morphology changes (detachment of the Small Sea, formation of the largely separated western and eastern basins), the continuing shallowing, and salinity build-up had crossed some critical threshold line, triggering major feedbacks and resulting in qualitatively different oceanographic behavior of the Sea. The key words characterizing this time interval are *vertical stratification* controlling the thermal regime, mixing and ventilation of the lake, and *interbasin exchanges* which have largely controlled the stratification through intrusions of denser water of the eastern basin into the western basin, and vice versa. The hydrography changes during this period have been accelerated and basin-specific, the salinity increase was non-linear in time, the SST regime with its seasonal cycling and the ice regime have changed significantly, and the evaporation rates have been increasing. At present, we are likely to be at the doorway to the third, yet further desiccation regime, which we discuss in Chapter 4.

4

The future: What happens next?

It is of obvious practical importance to predict the future changes of the Aral Sea level. The future developments, however, will depend on a number of "external" factors whose future behavior is hard to foresee. We have little means to predict the forthcoming interannual to decadal-scale climate changes modulating the river discharges into the Aral Sea and the evaporation from the lake surface. On the other hand, the forecast can be obtained in the measure that is determined by intrinsic dynamical properties of the system. Therefore, in this chapter, we do not aim at formulating any "deterministic" prediction—rather, we focus on analyzing the mechanisms that would favor one or another future scenario.

We must note that in the 1990s, the desiccation and salinization of the Aral Sea has progressed faster than it had been predicted in most previous prognostic studies. An example of such forecasts released by 1990 is shown in Figure 4.1 along with the factual data. The forecast shown in the figure was obtained through a probabilistic approach in which the water budget components entering Eq. (3.1) were treated as normally distributed random quantities, with the expected anthropogenic trends superimposed on them (Khomerini, 1969, 1978). This particular forecast is based on a specific scenario of river water diversions for irrigation in the 1990s, which turned out to be rather realistic. However, Bortnik and Chistyaeva (1990) gave a review of a number of other predictions that had been made to date for other irrigation scenarios.

Today, we know that in the 1990s, the factual lake level drop and salinity increase were larger than those predicted in most of the earlier forecasts by at least 2–3 m and 20–30 ppt, respectively. This may point to the existence of significant feedback mechanisms in the system which may not have been fully taken into account in the early predictions. For instance, many prognostic studies assumed constant mathematical expectation for the annual evaporation rates (Bortnik and Chistyaeva, 1990). However, the evaporation rates, as well as the other water budget

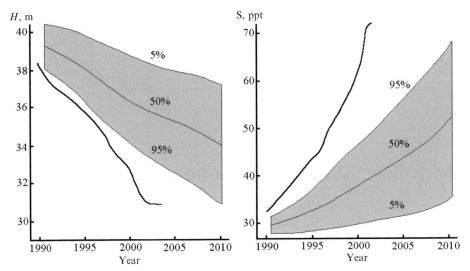

Figure 4.1. Changes of the Aral Sea level and salinity, as expected in 1990 for a likely irrigation scenario, and those really observed. The fine curves and the gray shading are the predictions and the respective confidence intervals (Bortnik and Chistyaeva, 1990), the bold curves are the factual observations (*cf.* Chapter 3).

components such as the groundwater inflow, for example, are likely to have changed in the course of the desiccation and as a result of it. Therefore, we begin this chapter with a discussion of the possible feedbacks that may at least partly determine the further changes of the Aral Sea characteristics.

4.1   FEEDBACKS IN THE SYSTEM

As expressed by Eqs (3.1) and (3.2), the temporal changes of the Aral Sea volume are fully determined by the water budget components (i.e., the evaporation, precipitation, river discharges, and groundwater inflow) on the one hand, and the hypsometry of the lake on the other. In turn, the budget components proper may be subject to change as a function of the lake volume, thus constituting feedbacks. This primarily refers to the evaporation rates which depend directly on the temperature and salinity regimes of the lake and, hence, are likely to have been modulated during the Aral Sea desiccation. But the Sea shallowing and shrinking could also have altered the exchanges with the groundwater table through changing the corresponding hydraulic gradients, changing precipitation by modifying the regional moisture budget, and, to some extent, even the river discharges by changing the "delta retention" conditions.

4.1.1   Evaporation ⇔ SST

There are strong indications that the net annual evaporation rates from the Aral Sea surface have increased by about 10–20% over the desiccation period, and the

maximum increase has occurred in the 1990s. First of all, this conclusion immediately follows from the water budget closure, even if the groundwater exchanges are
neglected (*cf.* Figure 3.6; see also Small et al. (2001a))—and if a significant groundwater inflow is to be allowed, this would imply a yet larger increase in evaporation.

A substantial increase in the evaporation rates was also reported by Small et al.
(2001a) who used a coupled regional climate lake model to simulate the hydrological
changes in the Aral Sea accompanying the desiccation. This instructive model study
will be repeatedly referred to in this section, so we first briefly identify the models.
The regional climate model was the National Center for Atmospheric Research
(NCAR) RegCM2 described elsewhere (e.g., Giorgi et al., 1993a, b)—a primitive
equation, σ vertical coordinate, grid-point limited area model with compressibility
and a radiative transfer scheme explicitly accounting for clouds, water vapor, and
carbon dioxide in the atmosphere (Briegleb, 1992). The model also includes an
explicit cloud water scheme that prognostically calculates the precipitation (Giorgi
and Shields, 1999). The lake model adapted for the Aral Sea and interactively
coupled with the regional climate model was a 1D model representing the vertical
heat transport by convection and turbulent mixing, with eddy diffusivity parameterized after Henderson–Sellers (1985). Details of this model can be found in Hostetler
and Bartlein (1990) and Hostetler et al. (1994). Evaporation in the model was
parameterized through the formula:

$$E = \rho_a C_D U_a (q_s - q_a) \tag{4.1}$$

where the subscripts a and s refer to air and the lake surface, respectively; ρ is the
density; C_D is the drag coefficient; U is the wind speed; and q is the specific humidity.
Of course, the form of this expression is similar to that of the bulk formula of Eq.
(2.1). The lake model was coupled with the regional climate model (whose domain
encompassed the entire Aral Sea drainage basin), so that at each lake model time
step (i.e., 30 minutes), the air temperature, surface pressure, humidity, wind speed,
precipitation, and effective radiation flux calculated in the atmospheric model were
passed to the lake model. The lake model was then used to simulate the Aral Sea sea
surface temperature (SST) and the evaporation from the lake surface. A series of
continuous 5.5-year-long simulations were completed, varying the spatial extent,
depth, and salinity of the Aral Sea in each experiment, aimed at examining the
effects of changes in the Aral Sea volume and hypsometry on the evaporation and
precipitation.

As reported by Small et al. (2001a), these model experiments showed that
following the twofold shrinking of the lake surface, the evaporation rates have
increased from April until September, with the largest increase of up to 15% in
July. In contrast, the evaporation rate has decreased between November and
February. However, the overall net annual evaporation has increased by about
5%. Furthermore, most of these changes in evaporation were due to changes in
the saturated water vapor pressure at the lake surface q_s entering Eq. (4.1), which
is mainly controlled by the SST. Therefore, the desiccation-related increase in the
summer SST is identified as the primary cause of the enhanced evaporation.
However, the observed summer SST increase is attributed to not only lake

shallowing but also to a larger scale warming across Central Asia throughout the desiccation interval, which may be a result of natural climate variability. The secondary cause of the Aral Sea evaporation enhancement is a decrease of q_a resulting from replacing a large portion of the original lake with desert which reduces the evaporation at the regional scale, thus lowering the specific humidity.

The essence of the $E \Leftrightarrow SST$ feedback mechanism is as follows. As lake shallowing progresses and mean depth decreases, the surplus of the solar irradiance falling on a unit area of the lake surface in summer is distributed over a smaller water column, thus leading to a larger increase of SST. Because q_s is an increasing function of SST, this in turn results in enhanced summer evaporation, *cf.* Eq. (4.1), and hence accelerated further shallowing of the lake, and so on.

Both the mechanical diminishing of the water body and the accompanying salinization play an important role in this mechanism, because saltier water generally accumulates in the bottom portion of the water column which elevates vertical stability and reduces vertical mixing. This leads to effective "trapping" of the excessive summer heat in the uppermost part of the lake and an additional increase of SST. This stratification effect was not taken into account in the model experiments by Small et al. (2001a) where the salinity was prescribed constant throughout the water column. Probably, it was for this reason that the simulated overall annual increase of the evaporation (5%) was smaller than that inferred from the water budget.

The increase in evaporation rates over the course of the desiccation can also be estimated directly from the observational q_a and SST data. As a proxy for the mean regional q_a we use the daily humidity data collected at a height of 2 m above the ground at the Aralsk meteorological station over the period 1958–1998, converted to the monthly averages (Figure 4.2). We calculate the seasonal cycle of q_2 using the data for the period before 1981 (i.e., the early desiccation stage), and then, separately, calculate the seasonal cycle using the data for the period after 1981 (i.e., the advanced desiccation stage). The curves shown in the Figure are obtained as sums of the annual and semi-annual Fourier terms best fitting the data. In other words, the curves represent functions in the form:

$$q(t) = q_0 + \sum_{i=1,2} [A_i \sin(i\omega t) + B_i \cos(i\omega t)] \tag{4.2}$$

where t is time, ω is the cyclic frequency corresponding to the annual period, q_0 is the overall average, and A_i and B_i are constants obtained by minimizing the r.m.s. deviation of the functions from the respective observational data. It can be seen that the characteristic values of q_2 and their seasonal cycles before and after 1981 practically coincide, with only a very slight decrease in the regional air humidity in summer and slight increase in winter (*cf.* Small et al., 2001a).

In contrast, significant desiccation-related changes in q_s can be established based on the observed SST changes discussed in Section 3.3.1. The saturated water vapor pressure at the Sea surface is fully determined by temperature T and salinity S. These dependencies for the World Ocean water (e.g., *Oceanographic Tables*, 1975) are depicted in Figure 4.3. Strictly speaking, in the case of the Aral Sea, they may be

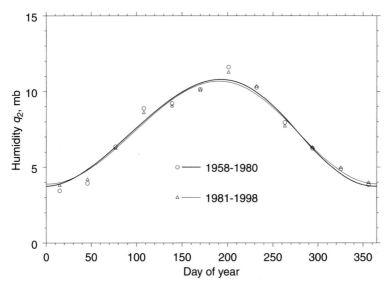

Figure 4.2. Seasonal cycle of near-surface humidity q_2 (mb) before and after 1981. The curves represent a sum of the annual and semi-annual Fourier terms best fitting the observational data. Aralsk meteorological station.

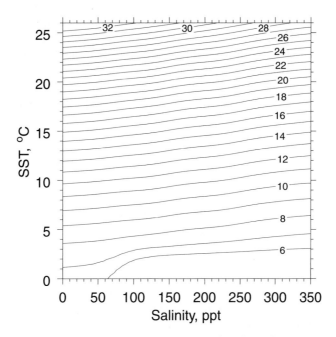

Figure 4.3. Saturated water vapor pressure (mb) as a function of water temperature and salinity.

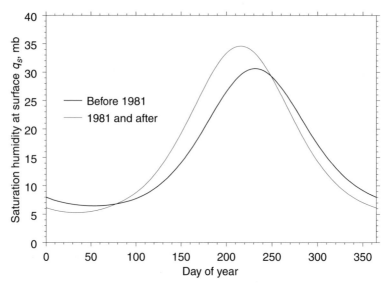

Figure 4.4. Seasonal cycle of saturated water vapor pressure q_s (mb) at the lake surface, before and after 1981. The curves represent sums of the annual and semi-annual Fourier terms calculated from the SST data (*cf.* Figure 3.8).

slightly different because of the different chemical composition, however, no better data is available. Under the assumption that the oceanic relations do hold in this case, we calculated the mean seasonal cycle of q_s before and after 1981 from the satellite-derived SST data shown in Figure 3.8 and the function $q_s\,(T,\,S)$ visualized in Figure 4.3. The result is shown in Figure 4.4. The maximum summer q_s at the advanced desiccation stage exceeds that in the early desiccation period by up to 5 mb, while in winter, the difference has the opposite sign and is up to 1–2 mb. There is also a notable phase shift in the seasonal cycle of q_s by about 20 days.

From the data presented above, we can now calculate the mean seasonal cycle of the term $(q_s - q_2)$ entering Eq. (2.1), or any other similar "bulk" formula which may be used to parameterize the evaporation from the Aral Sea surface. The cycles at the initial and advanced stages of the desiccation are shown in Figure 4.5. Once again, the curves represent sums of the corresponding annual and semi-annual harmonics. The gray shading indicates possible error intervals ($\pm$ r.m.s. deviation). The integral ratio between the two curves is about 1.11. This should constitute an 11% increase in the net annual evaporation rates, consistent with the modeling results by Small et al. (2001a) and the figures obtained from the water budget closure.

In summary, it can be said that the lake volume – SST – evaporation feedback discussed above is likely to have played an important role at the advanced stages of the desiccation, leading to a substantial acceleration of the shallowing progress in the 1990s. The manifestations of this mechanism should be best pronounced in the shallow eastern basin of the Large Aral Sea. A considerable part of the western basin is still deep enough so that its thermal regime could have remained largely

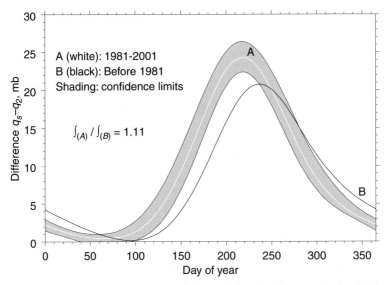

Figure 4.5. Mean seasonal cycle of $(q_s - q_2)$ in mb, before and after 1981.

unaltered. But the salinization of the lake has led to enhanced vertical stratification, which damped vertical mixing and may have resulted in additional elevation of summer SST within the western basin, thus reinforcing the feedback. As discussed in Chapter 3, the water originated from the hyperhaline eastern basin has been an important contributor to the western basin stratification. Therefore, the prospective effects of this feedback should depend on forthcoming morphology and salinity changes, especially those in the eastern basin, and the lifetime expectance for the channel connecting the western and the eastern basins.

However paradoxical it might sound, further desiccation will, at first, lead to a substantial *increase* of the mean depth of the lake, and its integral thermal behavior may partly return to that of a relatively deep water body in the near future. Indeed, the progressive level drop would result in the shrinking of the shallow eastern basin, so that the relative significance of the deep regions will increase. The mean depth h of the Aral Sea (defined as $h \equiv V/S$, where V and S are the lake volume and surface area, and calculated from the hypsometric relations given in Figure 2.3) is shown in Figure 4.6 as a function of the absolute lake surface elevation above the World Ocean level. Presently, the Aral Sea is at the minimum of the mean depth (~ 6 m). If the level decrease continues, the mean depth will rapidly increase until the absolute surface level is about 21 m a.o.l. The corresponding maximum value of h possible in the future is nearly 15 m, which is almost equal to the pre-desiccation mean depth. In addition, the drying up of the connecting channel and the separation of the eastern and the western basins of Large Aral should slow down the salinity and stratification build up in the western basin. Therefore, it seems reasonable to expect that the evaporation increase associated with the feedback mechanism discussed in this sub-section could cease in the near future.

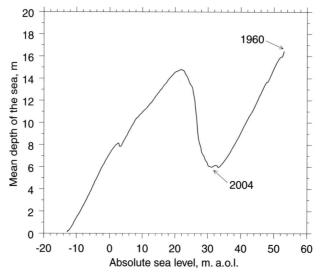

Figure 4.6. Mean depth of the Aral Sea as a function of the absolute sea surface level.

4.1.2   Evaporation ⇔ salinity

Another feedback mechanism tends to reduce the evaporation rates as the water salinity increases. Indeed, q_s entering Eqs (2.1) and (4.1) depends not only on SST, but also on the salinity (Figure 4.3). Of course, this is also the case for alternative parameterizations of the evaporation such as that by Calder and Neal (1984). For a fixed temperature, the saturation water vapor pressure at the surface, and hence the evaporation, are decreasing functions of the salinity. It is for this reason that, as is well known, evaporation from many brine lakes is reduced. This has made a number of earlier researchers hypothesize that in the course of the ongoing desiccation, the salinity growth effect on the evaporation should essentially compensate that from the increase of summer SST (Zaikov, 1952; Samoilenko, 1953). However, the salinity increase observed to date since the desiccation onset ($\Delta S \sim 80\,\mathrm{ppt}$ in the western basin, $\Delta S \sim 150\,\mathrm{ppt}$ in the eastern basin, $\Delta S \sim 10\,\mathrm{ppt}$ in the Small Sea) can hardly make up for the observed increase of summer SST ($\Delta T \sim 3\text{--}4°\mathrm{C}$). Indeed, as can be seen from Figure 4.3, even the maximum salinity increase observed in the eastern basin may have led to a maximum q_s drop only by about 2 mb, while the SST increase corresponds to an increase of q_s by 5–6 mb. The modulation of the evaporation rates by increasing salinity was investigated by Benduhn and Renard (2004) using a coupled numerical model of the water and salt budgets of the Aral Sea. They reported only a minor expected decrease of the evaporation due to this mechanism by approximately 4% from 1980 through to 2020.

Thus, at present, this negative feedback is likely to play a mere secondary role. Nonetheless, it might gain primary importance in the future when (and if) the salinity increase attains values of 200–300 ppt. The most likely candidate for such a scenario is the eastern basin after its separation from the remainder of the Large Sea. The salinity–evaporation feedback could be a major controller decelerating or eventually stopping further desiccation of the eastern basin, even if the river discharges into the basin are small.

4.1.3   Groundwater discharge ⇔ lake level

As discussed in Chapters 2 and 3, the lake–groundwater exchanges have been generally believed to be only a minor component of the Aral Sea's water and salt budgets. In more recent literature, however, it has been repeatedly argued that the relative role of the groundwater inflow may have greatly increased during the course of the desiccation (e.g., Stanev et al., 2004; Peneva et al., 2004), and some authors set their hopes of Aral Sea stabilization upon this possible feedback.

According to the classical Darcy's law (Darcy, 1856), the volumetric rate of a water flow through a porous or granular media is proportional to the applied hydraulic pressure gradient. In the case of our interest, this implies that the groundwater flow is essentially proportional to the groundwater table slope. Because the Aral Sea level drop must have resulted in a corresponding lowering of the groundwater table in the adjacent areas, this should indeed enhance the hydraulic gradients and increase the groundwater inflow to the Sea. This mechanism has been quantitatively investigated by Jarsjö and Destouni (2004). They used the following formula to assess the possible increase of the inflow:

$$\frac{G}{G_0} = \frac{X_{\text{bound}}(z_{\text{bound}} - d_{\text{gw}} - z_{\text{sea}})}{(X_{\text{bound}} + X_{\text{sea}})(z_{\text{bound}} - d_{\text{gw}} - z_{\text{sea},0})} \qquad (4.3)$$

where G is the specific groundwater discharge for the present conditions and G_0 is the specific groundwater discharge for the original, pre-desiccation conditions. In this formula, X_{bound} is the horizontal distance from the former Sea shore to a remote location where the Sea desiccation is no longer "felt" by the groundwater table, z_{bound} and d_{gw} are the respective absolute elevation and depth of the groundwater table at this remote location, z_{sea} and $z_{\text{sea},0}$ are the respective absolute elevations of the lake surface at the present and in the pre-desiccation state, and X_{sea} is the horizontal retreat of the Sea shore. The formula which follows immediately from Darcy's law is illustrated by the schematic in Figure 4.7. In the right-hand side of Eq. (4.3), X_{bound} and d_{gw} are unknowns, while the other variables can be estimated from observations.

About 10,000 plausible combinations of the parameters were analyzed by Jarsjö and Destouni (2004) to investigate their effect on the ratio G/G_0, considering the hydrogeological conditions in different parts of the Aral Sea. They concluded that although some increase of the groundwater inflow has indeed accompanied the ongoing lowering of the Aral Sea surface, this enhancement is likely to be moderate. For the eastern basin of the Large Aral, where the horizontal retreat of the Sea is at a maximum and the change of z_{sea} in the numerator of Eq. (4.3) is largely offset by the increase of X_{sea} in the denominator, the ratio G/G_0 is between

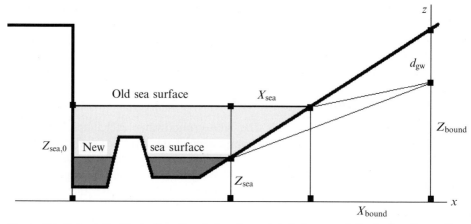

Figure 4.7. Schematic illustrating the desiccation-related change of the hydraulic gradient governing the groundwater inflow.

Adapted from Jarsjö and Destouni (2004).

0.9 and 1.2, indicating essentially unchanged groundwater flow conditions in this part of the lake. This result was quite robust with respect to changes of the parameters. In contrast, the ratio calculated for the western basin is between 1.2 and 1.5 for most of the plausible parameter combinations, and can even be up to 2 for some combinations. Finally, the results for the Small Sea turned out to be rather sensitive to the adopted value of the unknown variable X_{bound}, and the inferred present groundwater discharge into the Small Aral may be anywhere between nearly the pre-desiccation value and 200% of it.

According to some data, the groundwater component of the Aral Sea water budget is probably dominated by the deltaic aquifer, rather than the deep confined aquifer (Benduhn and Renard, 2004). Overall, the desiccation is likely to have led to a regional redistribution of the groundwater inflow between different parts of the Aral Sea, but only a modest net inflow increase. Given that, according to most of the available estimates, the pre-desiccation groundwater inflow rate was of the order of 1 km^3/year or less (see Section 2.3), the present values above only a few km^3/year do not seem very feasible. In our opinion, the hypothetical effect of the groundwater inflow increase on slowing down the Sea level drop alleged by some recent publications is open to discussion and should not be overestimated. Nonetheless, the negative "lake level ⇔ groundwater inflow" feedback may indeed play a considerable role at the advanced stages of the desiccation, especially in the deep western part of the Large Aral.

4.1.4   Other possible feedbacks

These are supposedly of secondary importance and will be discussed here only briefly. As is known, the Aral Sea shrinking has itself modified the regional climate to a certain extent. This in turn may have resulted in a reciprocal modulation of the Sea's water budget components (e.g., through changes in precipitation or evaporation rates depending on the air humidity, stability, and wind speed). These changes, however, are spatially confined to a limited area and often masked by the natural variability at different scales. For these reasons, and because the network of meteorological stations around the Aral Sea is sparse, the desiccation-related changes in the regional meteorology are underexplored from the observational standpoint (maybe except those in the air temperature, see Section 3.8).

In the above-cited model study by Small et al. (2001a), it was shown that the Sea desiccation should result in an increase of the precipitation rate over the lake surface by about 10 mm/year, or about 10%. Such a slim enhancement is largely offset by the natural interannual and seasonal variability and hardly observable. In any case, this increase would contribute only a few tenths of km^3/year at most into the overall water budget of the present Aral Sea, given that the precipitation is a minor component of the budget. Therefore, even if proven real, this possible feedback should be of minor significance.

According to Small et al. (2001a), the specific air humidity over the lake has decreased by up to 1 g/kg (over 1.5 mb) in the course of the desiccation because of the shrinking of the Aral Sea area and the corresponding reduction of the regional-scale evaporation. In the simulation, this effect was most pronounced in the summer and least pronounced in spring. In June, it accounted for up to 30% of the increase in the difference $(q_s - q_a)$ entering Eq. (4.1). If so, this mechanism alone could be responsible for an increase of the net summer evaporation by about 5% out of the observed 10–20%. But the observational data (Figure 4.2) suggest a much smaller factual drop in the specific humidity, and the decrease of q_a is very small compared with the increase of q_s. The effect of desiccation-related changes in C_D and U_a on the evaporation (Eq. 4.1) is believed to have been negligible (Small et al., 2001a).

Another possible feedback worth mentioning is related to the "delta retention" of the river discharges. It is more than likely that the related river inflow losses, formerly amounting to up to 8 km^3/year (Létolle and Mainguet, 1996), have changed significantly because of the Sea retreat from the former deltas. Some authors argue that the losses have actually decreased as the streams have become entrenched below the level of the deltas (e.g., Small et al., 2001a). On the other hand, satellite imagery indicates that the residual river flows often spread over considerable areas on the dry bottom. In any case, it seems obvious that the retention should increase as the lake limit retreats farther. This mechanism could play an important role modulating the future river discharges into the lake. Presently, Amu-Darya feeds the shallow eastern basin of the Large Aral, where the shoreline has already retreated far from the original delta. If the desiccation progresses, the eastern basin may soon greatly shrink in size, and the residual Amu-Darya runoff will have to make its way through up to 150–200 km of the dry former bottom terrain before it merges into the water body, which could lead to additional water losses at the advanced stages of the desiccation.

4.2   LIKELY SCENARIOS

The future dynamics of the Aral Sea level is fully determined by its water budget as expressed through Eqs (3.1) or (3.2). The total evaporation from the Aral Sea surface decreases as the lake area shrinks. Therefore, unless the income components of the budget are set to absolute zero, the Sea will stabilize sooner or later. Obviously, the stabilization condition is the equality of the total evaporation E on one hand, and the precipitation P, groundwater inflow G, and river discharges R on the other:

$$(E - P)S_e = (R + G) \tag{4.4}$$

where E and P are in m/year, R and G are in m^3/year, and S_e is the equilibrium surface area in m^2. It immediately follows from the latter equation that the stability conditions are graphically expressed by straight lines in the plane of the variables $(E - P)$ and $(R + G)$, and the slope of each line, symbolizes the corresponding equilibrium area. The equilibrium volume V_e of the lake, or its parts separated in course of the desiccation, can be found either by using Eq. (4.4) to obtain S and then converting it to V through the hypsometric relations, or otherwise numerically integrating Eq. (3.1) forward with the given parameters until the volume changes become negligible. Equation (3.2) can be used similarly to obtain the equilibrium mean depth of the lake. Generally, the variables E, P, R, and G are functions of time t. However, for the purposes of the simple diagnostic analysis presented below, we keep them constant with respect to t throughout each integration (these constant values can be intuitively interpreted as the averages over the integration period), but do vary them from one integration to another.

   The equilibrium volume and the time interval needed to approach the equilibrium state depend on the hypsometric relations for the lake (i.e., the relations connecting the lake volume, surface area, and the absolute surface elevation). The integral relations for the pre-desiccation Aral Sea were calculated in Chapter 2 from the bottom topography at about 1-km spatial resolution (see Figure 2.3). However, the integral relations are not of interest now, given that the Sea has already split into two parts (Large Sea and Small Sea), and the eastern and western parts of the Large Sea are at high risk of separating soon. The individual basins must be addressed separately. Therefore, we calculate the individual hypsometric relations for the Large Sea and its parts (the eastern and western basins taken together prior to their possible separation, the eastern basin alone, and the western basin alone), and the Small Sea. For this calculation based on direct integration of the bottom topography shown in Figure 2.2, we define the Small Sea as the part of the Aral Sea north of 46.2°N and east of 60°E, the western Large Aral as the part of the lake west of 59.5°E, and the eastern Large Aral as the remainder. The relations are shown in Figure 4.8. At present (2004), the western Large Aral area is 5,500 km^2 and its volume is 70 km^3. The figures for the eastern Large Aral are 10,500 km^2 and 25 km^3. The western and the eastern basins of the Large Aral Sea separate from each other when their respective volumes are about 60 km^3 and 15 km^3, and their respective areas are about 4,500 km^2 and 7,600 km^2. The present volume and area of the Small Sea are about 23 km^3 and 2,800 km^2, which is nearly equal to the values

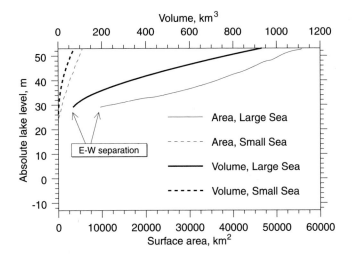

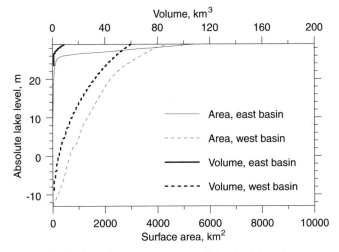

Figure 4.8. Hypsometric relations for separate parts of the Aral Sea. (*upper panel*) Small Sea and Large Sea (the eastern and western basins connected). (*lower panel*) Large Sea, the eastern and western basins disconnected.

corresponding to the late 1980s when the Small Sea detached from the main body of the lake.

Using the hypsometric relations, we integrate Eq. (3.1) numerically for a variety of plausible combinations of $(E - P)$ and $(R + G)$ to investigate how the equilibrium volume and the time needed to approach the equilibrium depend on the water budget components. In the numerical experiments, we consider that the "equilibrium" is established as soon as the interannual changes of the lake level become smaller than

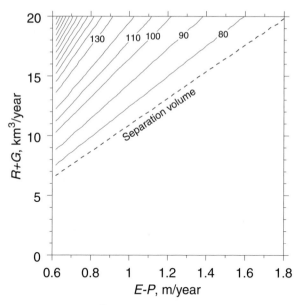

Figure 4.9. Equilibrium volume (km³) of Large Aral as a function of evaporation and river and groundwater inflow (eastern basin and western basins still connected).

0.25 m/year. We individually address the Small Sea and the two parts of the Large Sea before and after their separation. The results are presented in Figures 4.9–4.16.

4.2.1   Unseparated Large Sea

The expectations for the fate of the Large Aral while the eastern and the western basins are still connected are summarized in Figures 4.9 and 4.10. The basins will not separate in the future if the parameters are such that the corresponding domain on the $(E - P, R + G)$ plane lies above the dashed line in Figure 4.9, or equivalently:

$$R + G \geq 12.1(E - P) - 1.7 \tag{4.5}$$

(we remind the reader that the units for the river and groundwater inflows are km³/year, and those for the evaporation and precipitation are m/year.) As long as Eq. (4.5) holds, the Large Sea remains unseparated and eventually stabilizes at any volume above 75 km³. The most likely stabilization volume is, in this case, between 75 km³ and 120 km³, as these values result from the most plausible combinations of the water budget components. Such a volume corresponds to the absolute lake level standing between 29.7 m and 32.6 m a.o.l., the mean depth about 6 m, and the maximum depth between 42 and 45 m. The area of the deep western basin would remain essentially unchanged at around 6,000 km², and that of the shallow eastern part could be anywhere between 6,000 km² and 13,000 km² (i.e., 60–130% of its present value). The overall salinity should remain close to the present value.

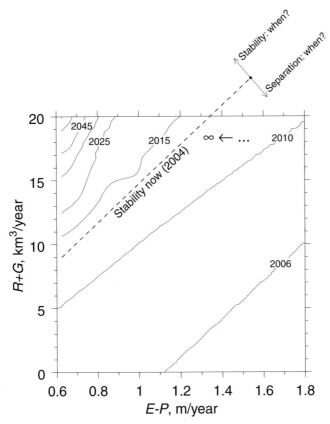

Figure 4.10. Expected time of the equilibrium or separation of the eastern and western basins, as a function of evaporation and river and groundwater inflow.

If the effective evaporation stays within its typical range approximately between 0.9 m/year and 1.2 m/year, this stabilization scenario would require annual river and/ or groundwater inflow of at least 9–12 km^3, which was indeed the case in 2002 and 2003. In the case where this enhanced inflow is maintained, the Large Aral could continue to remain unseparated and stabilize now or in the near future (Figure 4.10). In a marginal scenario where the river and groundwater inflows are elevated while the evaporation rates are reduced (upper left-hand corner of Figures 4.9 and 4.10), the Large Sea starts filling up considerably and attains a volume up to 200 km^3 by the mid-21st century. This, however, would require an increase of the river discharges up to 20 km^3/year and/or a decrease of the effective evaporation. None of the feedbacks discussed in the preceding section seem to hold promise for such changes, therefore, this optimistic scenario is unlikely.

If the water budget components do not satisfy inequality (4.5), then the western and the eastern parts of the Large Sea will separate in the future. This will definitely happen if the income components fall below 7 km^3/year, but even much higher

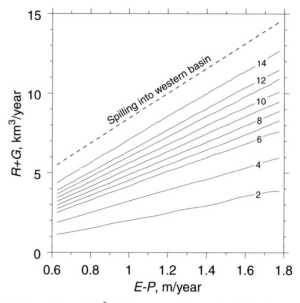

Figure 4.11. Equilibrium volume (km^3) of the eastern Large Aral after its possible separation, as a function of evaporation and river and groundwater inflow.

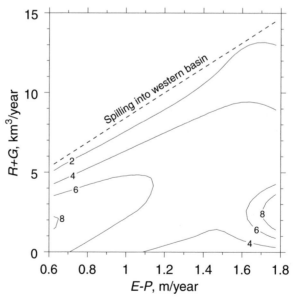

Figure 4.12. Estimated time (years) needed to achieve equilibrium for the eastern Large Aral after its possible separation, as a function of evaporation and river and groundwater inflow.

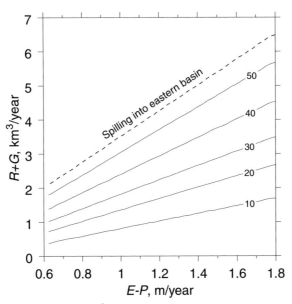

Figure 4.13. Equilibrium volume (km^3) of the western Large Aral after its possible separation, as a function of evaporation and river and groundwater inflow.

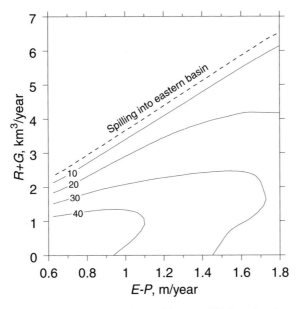

Figure 4.14. Estimated time (years) needed to achieve equilibrium for the western Large Aral after its possible separation, as a function of evaporation and river and ground water inflow.

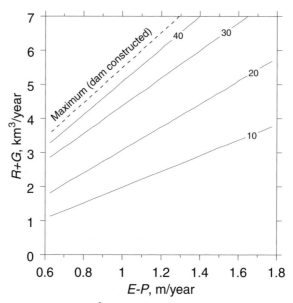

Figure 4.15. Equilibrium volume (km^3) of Small Aral as a function of evaporation and river and groundwater inflow.

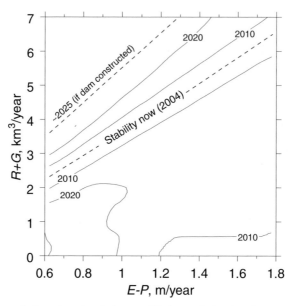

Figure 4.16. Estimated time (year) of the equilibrium of the Small Sea as a function of evaporation and river and groundwater inflow.

inflow rates may not be salutary if the effective evaporation is sufficiently large (see Figures 4.9 and 4.10). The separation may occur at any time, but the most likely period is between 2006 and 2010, which would correspond to the most plausible parameter combinations, as seen from Figure 4.10.

As soon as the Large Sea splits into two parts, their water budgets must be investigated on a separate basis.

4.2.2   Eastern Large Aral after its possible separation

Upon the separation, the remaining $15\,\mathrm{km}^3$ of the eastern basin water will be distributed over the broad area above $6{,}000\,\mathrm{km}^2$. The mean depth of the basin will be only about 1.8 m. Therefore, the basin will act much like a large evaporation pond. A subsequent increase of summer SSTs and, hence, evaporation rates should be expected (see Section 4.1). However, it must be kept in mind that after the separation, virtually all river discharges into the former Large Sea will be "trapped" in the eastern basin. On the other hand, the Amu-Darya runoff itself can be affected by the retreat of the basin from the former delta.

The stability diagrams for the eastern basin are presented in Figures 4.11 and 4.12. It is seen that the basin should shrink to an equilibrium state within only a few years following the separation. This result is quite robust with respect to the parameters. Complete vanishing of the eastern basin is theoretically also possible within this time frame if the inflow water budget components are set to zero, but this is an unlikely event, considering that at least some residual river discharges and groundwater input should be there and the salinity increase at the advanced stages of the desiccation should eventually result in a significant reduction of evaporation, see Section 4.1.2.

The residual volume, however, is highly sensitive to the river discharges and evaporation, and could be anywhere between nearly zero and $15\,\mathrm{km}^3$. For example, setting $(R+G)$ to $5\,\mathrm{km}^3$/year and $(E-P)$ to 1.4 m/year, which seems rather realistic in this case, yields the equilibrium volume of $6\,\mathrm{km}^3$. This would imply an area about $4{,}500\,\mathrm{km}^2$, a mean depth of 1.4 m, and a salinity increase up to around 300 ppt. On the other hand, maintaining $(R+G)$ above $10\,\mathrm{km}^3$ for a few years with the same effective evaporation would eventually lead to a spilling into the western basin. Such a spilling would occur for the domain:

$$R + G \geq 7.5(E - P) + 0.9 \qquad\qquad (4.6)$$

whose lower bound is shown by the dashed line in Figures 4.11 and 4.12.

In summary, the most likely scenario for the fate of the eastern basin after (and if) it separates from the western large Aral points towards its rapid shrinking into a residual brine lake whose equilibrium volume is only a few km^3 and whose mean depth is around 1 m. However, the volume, area, and salinity of the residual lake would be highly reactive to changes in the water budget components, especially the river discharges, and, therefore, subject to interannual and seasonal variability. Overspills into the western basin and reunifications of the Large Aral after the initial separation are also possible.

4.2.3   Western Large Aral after its possible separation

The stability conditions for the western Large Aral are illustrated by Figures 4.13 and 4.14. Virtually no river runoff will reach the western basin after (and if) the strait connecting it to the eastern part dries up. Therefore, the basin can only be eventually stabilized by the groundwater inflow. Because the surface area of the basin is relatively small, a modest groundwater discharge could equilibrate the water budget. As discussed in Section 4.1.3, the groundwater inflow into the western basin is expected to increase as the shallowing progresses. In addition, the evaporation rates in the basin can be expected to reduce following its separation (Section 4.1.1).

If the basin water budget components are such that:

$$R + G \geq 3.7(E - P) - 0.3 \tag{4.7}$$

then the equilibrium level is above the sill depth of the connecting channel, and the western and the eastern basins re-merge (dashed line in Figures 4.13 and 4.14).

For the most plausible range of the effective evaporation at 0.8–1 m/year, the annual groundwater discharge of $2 \, \text{km}^3$ would stabilize the basin at a volume of 40–$50 \, \text{km}^3$ within 10–20 years after the separation. Such an equilibrium volume corresponds to a surface elevation between 23 and 26 m a.o.l., a surface area between 2,700 and $3,500 \, \text{km}^2$, a mean depth of 14–15 m, and maximum depths of 37–40 m. Approximate estimates based on the salt budget point to an equilibrium salinity around 100 ppt or slightly higher.

However, if the groundwater discharge is at $1 \, \text{km}^3$/year or less, the stabilization may require a time frame over 40 years (see Figure 4.14) and the equilibrium volume would be below $30 \, \text{km}^3$.

4.2.4   Small Sea

The Small Aral Sea is a candidate for, at least, partial restoration. The Small Sea level has been relatively stable (with a variation of a few meters) since the separation from the Large Sea in the late 1980s (Micklin, 2004). With its present volume of $23 \, \text{km}^3$ and an average Syr-Darya inflow of 3–$4 \, \text{km}^3$/year, the Small Sea is close to equilibrium. Engineering plans are currently being implemented aimed at increasing the Syr-Darya discharge into the Small Sea by up to $4.5 \, \text{km}^3$/year (Micklin, 2004). This could be done if Syr-Darya's water periodically dumped into the Arnasay reservoir is diverted into the Small Sea. The project would also require that the flow between the Large Sea and the Small Sea is controlled by a new, structurally sound dam preventing the Small Sea water from overspilling into the Large Sea. The World Bank has approved the necessary funding and the work is reportedly underway (Pala, 2003; Micklin, 2004).

The stability conditions for the Small Sea are shown in Figures 4.15 and 4.16. The stabilization is achieved at the present level if:

$$R + G \sim 3.5(E - P) + 0.2 \tag{4.8}$$

which implies that the required inflow is between $2.8 \, \text{km}^3$ and $4.0 \, \text{km}^3$/year for the typical range of $(E - P)$. If the inflow is above these figures, the Small Sea volume will grow and could increase more than twofold by 2025 (provided that the dam is constructed). In this case, the lake level could exceed 46 m, the area would be nearly 5,000 km^2, and the salinity would return to the pre-desiccation value of about 10 ppt. On the other hand, in an unlikely but possible scenario where the discharges into the Small Sea are reduced for any reason, so that the left-hand side in Eq. (4.8) is significantly smaller than the right-hand side, the lake would shallow and then stabilize at a smaller volume (Figures 4.15 and 4.16).

4.3   CONCLUSIONS

It can be said that the Aral Sea, if taken as a whole, is rather close to dynamical equilibrium (*cf*. Benduhn and Renard, 2004). In fact, the lake has already been stable between 2002 and the present (2004), but this period was characterized by unusually high river discharges. If the annual river and groundwater runoffs into the Aral Sea persist at the rates characteristic for 2003–2004 (around 10 km^3 or higher), the Sea can remain relatively stable indefinitely, with the western and the eastern basins unseparated. We note that in a sense, the shallow, flat bottom eastern basin connected with the deep western region acts as a stabilizing controller itself: a small drop in the Aral Sea level results in a great decrease of the eastern basin area and, hence, corresponding reduction of the total evaporation from the lake, and vice versa.

However, if the discharges return to smaller values typical for the 1980s, the deep western and the shallow eastern parts of the Large Aral may separate in the near future. In that case, the future expectations vary for different parts of the lake.

Following the separation, the western basin is likely to continue shallowing, but at significantly slower rates, until it eventually stabilizes in the more or less distant future. The expected slowing down of the basin desiccation is attributed to moderate summer SSTs in the deep basin and the increase of groundwater discharges.

The future fate of the eastern basin is perhaps the least predictable, because its equilibrium volume is very sensitive to the water budget components, and anything between a complete drying of the basin over only a few years to its re-merger with the western basin is still possible within the plausible range of the river discharges and evaporation. Still, the most likely scenario is a rapid shrinking of the basin into a smaller terminal hyperhaline lake. The evaporation from the residual lake could be significantly damped by the increase of salinity.

Unless the Syr-Darya discharges into the Small Sea strongly reduce, the Small Sea will continue to be relatively stable at its present, or a larger, volume. If the planned water management and engineering measures are successfully implemented, the Small Sea could return to its pre-desiccation low salinity by the 2020s.

5

The Aral crisis in global perspective

The desiccation of the Aral Sea is considered to be among the worst disasters of its kind on record. However, the global list of water bodies experiencing significant desiccation, or otherwise endangered because of either unsustainable anthropogenic pressures or global climate change, is long. The negative consequences are manifold, ranging from deterioration of environmental conditions (desertification processes, increase of climate continentality) to economical and social impacts (decay of fisheries, tourism, and other related businesses). Because the Aral Sea represents an extreme case of lake degradation, insight obtained from Aral may have a broader applicability to other water bodies.

In Section 5.1, we discuss some such "dying" or endangered inland water bodies whose conditions are kindred to those of the Aral Sea in some respects. Of course, these cases are not exhaustive—small and medium size lakes suffering severe alterations of their regime because of anthropogenic activities in their catchment areas are virtually innumerable. We restrict this brief review to only a few illustrative selected examples best represented in the literature. A more detailed account can be found elsewhere (in particular, see the articles collected in the book edited by Nihoul et al. (2004)—a part of the review below is based on the materials published in this volume).

The extreme oceanographic environment of the present Aral Sea is also a good proxy for studying some physical and chemical processes that do occur (although manifested less dramatically) in other seas and the oceans, in particular, basins with strongly elevated salinity and/or stratification. Much like the "normal" seas, the Aral Sea has distinct water masses, thermohaline and wind-driven circulations organized in the form of gyres and eddies, and strong vertical and horizontal inhomogeneity. It also has separate basins exchanging their waters and properties through the connecting strait. But, in contrast with the seas and oceans, the Aral Sea is compact in size, and while the absolute magnitude of the thermohaline variability characteristic for the Aral Sea can be comparable with that seen in the ocean, the corresponding

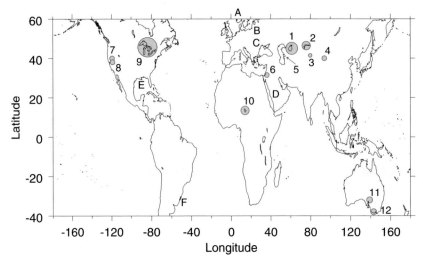

Figure 5.1. Geographic objects mentioned in the text: 1—Aral Sea; 2—Lake Balkhash; 3—
Lake Issyk–Kul; 4—Lake Lobnor; 5—Kara-Bogaz-Gol Bay and the Caspian Sea; 6—Dead
Sea; 7—Pyramid Lake; 8—Mono Lake; 9—Great Lakes; 10—Lake Chad; 11—Lake Eyre;
12—Lake Corangamite; A—Fjords of the Norwegian Sea; B—Baltic Sea; C—Black Sea; D—
Red Sea; E—Gulf of Mexico and the Mississippi delta; and F—Southern Brazilian shelf and
Plata estuary.

spatial scales are smaller by several orders of magnitude. For example, the density
difference between the surface and the bottom (depth about 40 m) water in the
western Aral Sea is nearly the same as that typical for the World Ocean (depth
about 4,000 m on average). In the present Aral Sea, we encounter a number of
oceanographic conditions that normally can only be seen in a laboratory experiment.
This makes Aral a "natural model", potentially useful for investigating some general
oceanographic processes and features, especially those linked with enhanced vertical
stratification and elevated salinity. Some of the related aspects are discussed in
Section 5.2.

A map of the locations referred to in this chapter is shown in Figure 5.1.

5.1　THE ARAL SEA AMONG OTHER CRITICAL LAKES

5.1.1　The Dead Sea

The Dead Sea is a deep terminal lake at the border between Israel and Jordan
(Figure 5.1). The present Dead Sea surface is located at about 416 m below the
World Ocean level, which makes the lake the lowest land spot on Earth (e.g.,
Gavrieli and Oren, 2004). The Dead Sea whose maximum salinity is above 340 g/l
and density is about 1,237 g/m^3 is considered to be one of the saltiest lakes in the
world.

The Dead Sea surface level has dropped by about 21 m since the mid-20th century. We note that this shallowing was almost simultaneous with the Aral Sea desiccation and, in absolute terms; the level drop was nearly equal to that observed in the Aral Sea. It is also notable that the volume of the Dead Sea is comparable with the volume of the present Aral Sea, but the surface area is much smaller, namely, about 625 km^2, therefore, the total evaporation is also smaller and only a relatively small increase of river inflow is needed to achieve the equilibrium water balance. Presently, the estimated annual deficit of the Dead Sea water budget is about 0.6 km^3 (Gavrieli and Oren, 2004).

The Dead Sea desiccation continues at rates of 0.5–1 m/year. The shallowing is believed to have been anthropogenic and resulted from major water management interventions in the drainage basin, manifested mainly through water diversions from the Jordan River feeding the lake. The river waters have been diverted for agricultural and industrial uses by Jordan, Syria, and Israel, and the discharge into the Dead Sea has reduced from 1.5 km^3/year in the 1950s to only 0.15 km^3/year at present (e.g., Al Weshah, 2000). Potash industries at the Dead Sea are also responsible for an estimated 30–40 cm of the annual level drop. These industries consume 0.2–0.3 km^3 of Dead Sea volume per year by diverting a considerable amount of water into evaporation ponds and returning only a part of it as much saltier "end brine" (Gavrieli and Oren, 2004). On the other hand, the lake volume drop is believed to have been slowed down by an increase of the groundwater inflow into the lake because of the hydraulic gradient enhancement through the mechanism discussed in Section 4.1.3 (note the analogy with the Aral Sea).

At present, the Dead Sea is 310 m deep. Morphologically, the lake used to consist of two basins, the large and deep northern part, and smaller shallow southern part. The two basins were separated by a peninsula and connected through a narrow strait near the western coast, which is again reminiscent of the present Aral Sea. The southern basin dried completely by 1977, except the areas occupied by the evaporation ponds.

Until 1979, the Dead Sea had been meromictic[1] for a long time (virtually, for centuries). The shallow southern part of the lake was still filled with water. The deep northern basin exhibited strong haline and density stratification, with the salinity increasing from around 300 g/l in the upper layer, which was typically about 40 m deep, to about 332 g/l near the bottom (Neev and Emery, 1967). Such a stratification originated essentially from the assimilation of river discharges in the surface layer (e.g., Ivanov et al., 2002) and had largely prevented any vertical mixing between the upper and the lower water masses. The bottom layer, therefore, was anoxic and contained sulphide (e.g., Gavrieli and Oren, 2004). This bottom water mass which had been isolated for a long period of time was sometimes called "fossil water" (Steinhorn et al., 1979). Following the increase of the anthropogenic diversions of the river water and the desiccation progress, the vertical density stratification eventually relaxed, which led to a major overturning event in February, 1979

---

[1] Meromictic (lake) = permanently stratified, usually without oxygen in its deeper portions, due to a density gradient and a lack of overturn.

Table 5.1. Content of major anions and cations in the Dead Sea water in summer 2002.
From Gavrieli and Oren (2004).

Ion	Cl^-	SO_4^{2-}	Br^-	HCO_3^-	Na^+	Mg^{2+}	K^+	Ca^{2+}
Content (g/l)	229	0.4	5	0.3	34	47	8	18

(Steinhorn et al., 1979; Beyth, 1980; Steinhorn and Gat, 1983). As a result, the vertical thermohaline structure became uniform. Since then, the lake has been normally holomictic,[2] with autumn and winter convection mixing and ventilating the water column, except during intermittent, relatively short meromictic periods (1980–1981, 1992–1994) of unusually rainy conditions and elevated river discharges, accompanied by a temporary rise of the lake surface level by 1–2 m and a surface salinity drop by up to 30% (Gavrieli and Oren, 2004). In the holomictic regime, stable density stratification in summer is controlled by a thermocline where the temperature decreases from up to 36°C in and immediately below the mixed layer to only about 22°C at the bottom. The temperature drop in the vertical profile is sufficiently large to offset the upper layer salinity increase due to enhanced summer evaporation. In autumn, cooling leads to a relaxation of thermal stratification and an overturning of the water column. The seasonal cycle of salinity and temperature has been modulated by a considerable general positive trend over the last decades (e.g., Anati, 1997; Anati, 1998).

The salt composition of the Dead Sea water (Table 5.1) is rather peculiar and significantly different from the composition of the Aral Sea (cf. Table 3.4).

It can be seen that the sulphate/chloride ratio SO_4/Cl for the Dead Sea is smaller than that for the Aral Sea by a factor of about 450. The Dead Sea water has a Ca-chloride type composition (i.e., the content of Ca^{2+} is larger than the content of SO_4^{2-} plus HCO_3^-). As in the Aral Sea, the precipitation of compounds from the oversaturated water has played an important role in the chemical regime of the Dead Sea. At present, the lake is saturated with respect to halite NaCl, aragonite $CaCO_3$, and anhydrite $CaSO_4$ (Gavrieli et al., 1989). In the course of the salinization, halite and gypsum $CaSO_4 \cdot 2H_2O$ have precipitated massively since 1982 until recently (Steinhorn, 1983; Gavrieli, 1997). During the last few years, the precipitation was relatively small, possibly because of the depletion of sulphate and bicarbonate associated with the decrease of river inflow. Nonetheless, the precipitation of halite has already resulted in a considerable change in the ion composition, in particular, the molar ratio Na/Cl has decreased by about 20%, while the ratio Mg/K has increased by about 10%, since the 1960s (Gavrieli and Oren, 2004).

From the biological point of view, the lake is not exactly dead, as a number of microbial communities live in the Dead Sea, despite its extremely high salinity. The dominant types are autotrophic, unicellular algae *Dunaliella* sp. and heterotrophic prokaryotes such as halophilic red Archaea of the family *Halobacteriaceae* (e.g., Oren, 1999; Gavrieli and Oren, 2004). Algal blooms normally occur under meromic-

---

[2] Holomictic = mixes completely throughout the water column at least once a year, or more frequently.

tic conditions, when the upper layer salinity drops by 10–20%, following winters with enhanced freshwater discharges. In the spring of 1992, *Dunaliella* populations as dense as 1.5×10^4 cells/ml were documented (Oren, 1993). On the other hand, little or no algae have been observed during the monomictic[3] periods. The red halophilic Arachaea rapidly grow coincident with the algal booms, thriving on the organic matter produced by the algae. At the peak of the archaeal bloom of 1992, the population density was up to 3.5×10^7 cells/ml (Gavrieli and Oren, 2004).

According to most of the model forecasts, the Dead Sea level will continue to drop for some time, until an equilibrium state is achieved (e.g., Yechieli et al., 1998). Because of the expected reduction of evaporation following the salinity increase in course of the progressive desiccation and highly hygroscopic nature of the Dead Sea solute, it is believed that the lake will never dry out completely, even if the river inflow is set to zero (Krumgalz et al., 2000). The steady-state level, however, depends on the volume of the inflow. If the present rates of the river discharge persist, the Dead Sea level is expected to drop another 150 m before the lake achieves steady state in about 200 years from now. These figures could be much more optimistic if the water supply from the Jordan River increases up to nearly its original values (Yechieli et al., 1998), but this scenario is unlikely in the foreseeable future. The "Peace Conduit" project which implies pumping the Red Sea water into the Dead Sea is being widely discussed now. If implemented, the plan could help to slow down the decline or stabilize the lake. However, all possible consequences of such an intervention are yet to be investigated and analyzed.

5.1.2   Lake Chad

The Lake Chad area has shrunk to nearly one-twentieth of its former extent, from approximately 25,000 km^2 in 1963 to 1,350 km^2 in 2001 (Kostianoy et al., 2004). Located in central Africa, the lake washes the territories of Nigeria, Niger, Chad, and Cameroon (Figure 5.1). Before 1960, the rivers feeding the lake supplied about 42 km^3 of water per year, on average (Lemoalle, 2004). Nearly 40 km^3 of this volume originated from the principal tributary, the Chari River merging into the southern part of the lake, while the two other sources of the inflow, rivers El Beid and Yobe, accounted for about 2 km^3/year. The river discharges, however, have been highly variable at the seasonal and interannual scales. Since the early 1960s, the inflow started to decrease at a rate of about 1 km^3/year^2, until the mid-1980s when the inflow was only about 10 km^3/year or slightly above. Since then, a moderate increase of the discharges has been observed, and the values between 20 and 30 km^3/year are characteristic of the present-day. We note that the pre-desiccation discharges and the patterns of their decrease since the 1960s are similar to the corresponding figures for the Aral Sea.

There has been controversy around the relative roles of the natural climate variability and anthropogenic factors in the Lake Chad desiccation. Some authors

---

[3] Monomictic = mixes completely throughout the water column once a year, in fall or winter.

have argued that about 50% of the lake shrinkage since 1960 should be attributed to anthropogenic water diversions which have increased fourfold between 1983 and 1994 (e.g., Kostianoy et al., 2004, and the literature cited therein). On the other hand, some quantitative estimates (e.g., Lemoalle, 2004) suggested that the total water withdrawal for irrigation in the Lake Chad basin actually does not exceed $0.2 \, \text{km}^3/\text{year}$, which is only about 1% of the observed drop in the river discharges. If so, the human induced factors have not contributed very significantly to the lake's water budget (Coe and Foley, 2001). It is generally thought that the main reason for the ongoing desiccation was the decrease of rainfall in the lake catchment area, in particular, along the Chari basin. A period of elevated precipitation over the West African Sahel starting from 1950 was followed by the period of low rainfall from the late 1960s through to the present-day. During this period, the mean annual precipitation sums decreased by about 150 mm, or 50% (L'Hôte et al., 2001; Lemoalle, 2004).

In the early 1960s, the absolute level of Lake Chad's surface was about 283 m a.o.l., subject to considerable spatial variability (± 0.4 m) depending on the wind conditions (Talling and Lemoalle, 1998). The present level of the remainder of the lake is below 279 m. It is well known that on a long temporal scale, Lake Chad, as well as the Aral Sea, has undergone several similar regression and subsequent transgression episodes in the past. The largest of the transgressions occurred between 12,000 and 6,000 years BP, when the surface area of the lake is believed to have been as large as $250,000 \, \text{km}^2$. In contrast, a very strong regression took place in the 15th century. At a shorter temporal scale, notable regressions have been documented around 1850 and between 1904 and 1915 (Lemoalle, 2004). Because of this energetic variability of the lake size and the continuous succession of dry and wet conditions, a classification into three archetypic states of the water body, namely the Large Lake Chad, the Small Lake Chad, and the Normal Lake Chad has been adopted in the related literature (Tilho, 1928). The Large Chad refers to lake level standings above 282 m, or a lake surface area exceeding $22,000 \, \text{km}^2$. The Small Chad conditions are those with a level under 280 m, which implies an area of $14,000 \, \text{km}^2$ or smaller. The Normal Chad is the intermediate case. Thus, the contemporary desiccation is identified as a transition from Large Lake Chad to Small Lake Chad.

The Normal Chad is naturally divided in two parts, namely the northern and southern basins, separated by a shallow and narrow transversal swell commonly referred to as the Great Barrier. Much like the Large Aral and Small Aral, the two basins became disconnected from each other in March, 1973. Separated from the main source of water supply, the Chari River whose mouth is at the southern extremity of the lake, the northern basin then dried up completely by 1975 and has been essentially dry since then, except for a part of it during a number of intermittent flood episodes followed by spillings from the southern basin through the Great Barrier (Lemoalle, 1991). These spillings and the related interbasin exchanges have apparently played a significant role in the regime of the surviving southern basin and the state of the biological communities in either basin. As argued by Lemoalle (2004), the division of a lake into separate smaller basins over the course of desiccation appears to be a feature common to many other critical water bodies (including

Table 5.2. Desiccation characteristics for the Aral Sea, Dead Sea, and Lake Chad. River discharge drop is an approximate difference between the characteristic pre-desiccation and present-day inflow.

	Period	Causes	River discharge drop (km³/year)	River discharge drop (%)	Lake level drop (m)	Area loss (10³ km²)	Area loss (%)
Aral Sea	1961–present	Anthropogenic + natural	~40	~80	23	50	75
Dead Sea	1900s–present	Mainly anthropogenic	~1	~90	21	0.3	35
Lake Chad	1963–present	Mainly natural	~20	~50	4	22	90

the Aral Sea), which may be interpreted as a mechanism to stabilize the level and maintain tolerable salinity in some isolated parts of the lake when the net evaporation over the entire water body cannot be offset by the net inflow.

The shrinking of Lake Chad has had strong impacts on its biological communities, ranging from plankton to fishes and birds, and led to a series of modifications in the natural environment, especially the marshy areas in the southern basin and on the Great Barrier. The biological and environmental consequences of the desiccation have been described in detail elsewhere (Carmouze et al., 1983; Lemoalle, 2004). The local economy has been affected significantly, but the damage was partly offset because the fisheries promptly switched to the newly dominant species and some areas of the former bottom have been used for agricultural needs (e.g., Sarch and Birkett, 2000).

The water budget analysis indicates that the lake has been close to equilibrium since its separation into different parts in the mid-1970s. Recent publications (Lemoalle, 2004) emphasized that the present Lake Chad is relatively stable as "Small Chad", rather than desiccating or "dying", although the state of the lake depends strongly on the annual river discharges. According to Lemoalle (2004), Chari runoff above 12 km³/year would preserve the status quo indefinitely. The annual discharge of 38–40 km³/year is needed to recover the pre-desiccation state (i.e., Normal Lake Chad). Such river runoffs could be achieved if the regional precipitation rates eventually increase again because of climate change, or otherwise if energetic water management measures are taken. For example, the possibility to transfer up to 40 km³ of water per year from the Zaire–Ubangui basin into the Chad basin is being investigated (Lemoalle, 2004). If implemented, this project would lead to rapid recovery of the Normal to Large Lake Chad.

The Lake Chad, the Aral Sea, and the Dead Sea represent, perhaps, the most well-known examples of the desiccation of large inland water bodies. As discussed above, there are both notable similarities and differences between the three cases. Some comparative characteristics are given in Table 5.2.

5.1.3 Other examples

We now mention only briefly some other illustrative examples of desiccating lakes all over the world, as reported in the literature.

Many of the desiccating or endangered inland water bodies are located in Asia. The arid or semi-arid zones of Asia, which are particularly vulnerable to changes in water balance, have been significantly affected by both regional climate change and unsustainable irrigation. The most striking (although not the most well documented in the literature) instance is Lake Lobnor in northern China (Figure 5.1), which vanished completely in 1972 (e.g., Kostianoy et al., 2004). The former lake bed was turned into a weapon testing site, raising additional ecological concerns. Some other lakes in the region have also been experiencing desiccation since the 1950s. The annual precipitation has reportedly decreased by at least 30% over this period. Consequently, Lake Ohlin at the head of the Yellow River has been shallowing at the rate of over 2 cm/year (Wang, 1993), while the level of the Lake Qinghai Hu has been dropping at the average of 10 cm/year between 1959 and 1982 and 1990 through to 2000s. The total level drop since 1908 is almost 12 m, and the salinity increased twofold from 6–12 g/l (Kostianoy et al., 2004). Another large salty lake in north-western China, the Ebinur Lake, has shrunk by 60%, from over 1,200 km^2 in the 1950s to 530 km^2 at present, which has resulted in the extinction of several plant and animal species in and around the lake. Considerable wind transport of dust from the former lake bottom has been estimated at a rate of up to 5 million tonnes per year. To minimize this harmful process, a project focused on planting trees around the lake and on the former bottom has been reportedly implemented (Kostianoy et al., 2004). As known, similar measures have been attempted on the newly dry bottom of the Aral Sea.

Located in the same latitude belt as the Aral Sea, another major lake of Central Asia—Lake Balkhash—neighbors Aral to the east (Figure 5.1). The lake whose area is 18,200 km^2 and whose volume is 105 km^3 (the parameters are very close to those of the present Aral Sea) is peculiar: its eastern part is brackish (salinity up to 7 g/l), while the western part, fed by the Ili River and largely separated from the remainder of the lake by the Sary–Isek peninsula, is fresh or nearly fresh (<1.5 g/l). The lake level and regime has been long known to depend strongly on the river discharges, whose principal constituent is the Ili runoff totaling 15 km^3/year, or 80% of the total freshwater inflow, on the long-term average. During the past century, the lake surface area and volume varied from 15,700 km^2 and 83 km^3 (1946) to 23,500 km^2 and 164 km^3 (1910) respectively. Since 1970, the Ili runoff has been reduced because of damming of the river and construction of a Kapchagay hydropower plant and reservoir (e.g., Shaporenko, 1995). Satellite altimetry data for the last 10 years indicated a progressive drop of 1.4 m in the lake surface level (Kostianoy et al., 2004). Lake Balkhash has also been facing serious challenges to its biological systems and fisheries, which, however, have been partly parried by efficient management (Petr, 1992).

Lake Issyk–Kul, south of Lake Balkhash (Figure 5.1), has also been facing a progressive drop of the water level by more than 3 m since the 1920s, along with increasing pollution of the lake (Giralt et al., 2004). There is no agreement among the specialists on whether the shallowing was caused by a long-term climate change, tectonic activity in the region, or anthropogenic water consumption for agricultural use (Romanovsky, 2002).

Another Central Asian water body that is worth mentioning in this context is Kara-Bogaz-Gol Bay, a large (about 20,000 km^2), shallow, hyperhaline lagoon east of the Caspian Sea, connected to it through a narrow strait (Figure 5.1). Generally, the Kara-Bogaz-Gol surface level is lower than that in the adjacent portion of the Caspian Sea, so the water flows into the bay where much of it evaporates. The bay is one of the saltiest bodies of water in the world: salt concentrations nearly as high as those in the Dead Sea (i.e., up to 350 g/l) have been documented. In 1980, the connecting channel between Kara-Bogaz-Gol was dammed. In response, the bay dried up almost completely by late 1983. After the dam was destroyed in 1992, the lagoon surface level increased at the rate of 1.5 m/year until it reached a relatively steady state by 1997. Recent satellite altimetry findings showed a slight trend towards shallowing at about 6 cm/year (Kostianoy et al., 2004), which is also characteristic for the Caspian Sea itself. As known, there were serious concerns around the Caspian level dynamics and water budget, which has been shown to be tightly connected with the discharges from the Volga River accounting for about 80% of the budget's income (e.g., Kosarev and Yablonskaya, 1994). Of course, the Caspian Sea with its total volume of 78,600 km^3 and maximum depth of 1,025 m is by no means at risk of desiccation. However, during the recent regression (1930s–1977), the sea level decreased by nearly 3 m attaining its historical record low in the last 400 years, which has had considerable negative consequences, especially in its biologically productive shallow areas. At present, the Caspian Sea level is almost at its pre-regression value (around 26 m below the World Ocean level), but the Sea is facing serious anthropogenic problems of a different nature, including, but not limited to, those connected with the explosive invasion of ctenophore jellyfish *Mnemiopsis leidyi*, threatening to destroy the food chains and biological communities of the Sea (e.g., Vinogradov et al., 2002).

A number of lakes in the Americas have also experienced desiccation in the past century. For instance, the deepest terminal salty lake of the western hemisphere, the Pyramid Lake in Nevada (Figure 5.1), has exhibited a notable level drop of more than 20 m since 1910 (Wheeler, 1974; Kostianoy et al., 2004). The present lake volume is 27 km^3, its area is 453 km^2, and its maximum depth is 101 m [*World Lakes Database*, 2004]. The shallowing was accompanied by a salinity increase from 3.8–5.5 g/l. The lake level migrations are associated with the variability of the discharges from the Truckee River, the main tributary of the Pyramid Lake. Another North American salty lake, the Mono Lake in California (Figure 5.1), has undergone a shallowing of about 17 m between 1920 and 1982 because of diversions of water from Mono's tributaries which began in 1941.

Since 1982, the lake level has recovered by a few meters. At present, the lake volume is about $3\,km^3$, the area is $180\,km^2$, and the maximum depth is $43\,m$. The present salinity is nearly $80\,g/l$ [*World Lakes Database*, 2004], and at the lowest level in 1982, it was $99\,g/l$.

The two lakes mentioned above are mere examples of desiccating or endangered water bodies on the continent, which are numerous. Even the world's largest fresh-water system, the Great Lakes, appear to be shrinking. In 2002, the aggregate level of the five Great Lakes (Superior, Huron, Michigan, Erie, and Ontario) was the lowest in more than 30 years (Kostianoy et al., 2004).

The surface level of Lake Corangamite, Australia (Figure 5.1), dropped by over $4\,m$ from $118\,m$ a.o.l. in 1960 to $113.8\,m$ a.o.l. in 2003 (Timms, 2004). The lake whose present volume is about $1\,km^3$ and maximum depth about $3\,m$ is considered to be the largest, permanent, salty lake in Australia. The salinity of Lake Corangamite has fluctuated in a broad range from below $10\,g/l$ to over $120\,g/l$, closely correlated with the local rainfall (Williams, 1986). However, in the last few decades, the mineraliza-tion has been constantly building up, due to diversions of water from the main tributary, the Woady Yaloak responsible for about 35% of the total inflow, into the Barwon River (Williams, 1995). The average salinity increased from about $25\,g/l$ in 1960 to $110\,g/l$ in 2002 (Timms, 2004). The desiccation and salinization had severely negative effects on the lake's biodiversity (e.g., the number of waterbird species populating the lake decreased from 36 in the late 1970s to 18 in the early 2000s). No fish or mycrophytes have recently been encountered, and the diversity of benthic communities has been greatly damaged. Presently, the Woady Yaloak diversion scheme is being reviewed by the local authorities. A solution is being sought which would allow a balance between the interests of users of land around the lake and the normal physical and chemical regimes of the lake. The target salinity is about $25\,g/l$, implying an increase in the level by $2-3\,m$ with respect to the present standing. Should this low salinity be achieved, it is hoped that the lake can fully recover its original biological systems.

Another notable example in Australia is Lake Eyre (Figure 5.1). Like the present Large Aral Sea, Lake Eyre consists of two separate parts, Eyre North ($140\,km$ long and $80\,km$ wide) and Eyre South ($65\,km$ long and $25\,km$ wide), connected through a narrow strait called the Goyder Channel. Because the lake's drainage basin is as large as $1,140,000\,km^2$, the lake level is very sensitive to even a small variability in rainfall. Located in one of the driest regions of the country, the lake is "episodic": it is usually dry, but during flood events, Lake Eyre temporarily becomes the largest lake on the continent, with a surface area up to $9,500\,km^2$. Strong transgressions occurred in 1950, 1974, and 1984, for example (Kostianoy et al., 2004), followed by desiccation periods. During the flood events, water and salt exchanges between the northern and the southern parts through the channel played an important role. For example, an estimated 30 million tonnes of salt were transferred from Lake Eyre North to Lake Eyre South during the transgression of 1974 (*cf.* Chapter 3 where we obtained 70 million tonnes as the respective estimate for the annual salt exchange between the western and eastern basins of the Large Aral).

We have tried to demonstrate in this section that the Aral Sea desiccation can be thought of as striking and illustrative but, unfortunately, not at all a unique manifestation of a worldwide process. Significant desiccation, salinization, and water quality deterioration have taken place in many inland water bodies over the past decades, often triggering serious environmental, ecological, economical, and health consequences. In 1986, the International Lake Environment Committee (ILEC) initiated a global "Survey of the State of World Lakes", encompassing 217 lakes (64 in Asia, 20 in Africa, 73 in North and South America, 56 in Europe, and 4 in Oceania) (Kira, 1997). The results have indicated that various environmental disruptions have been rather common for many lakes in all continents. In many cases, the degradation of lacustrine systems can be deservedly attributed to human impacts, in particular, diversions of water from tributary rivers. However, recurrent transgressions and regressions of inland water bodies are also caused by natural climate variability, or a combination of the anthropogenic and climatic factors. As Timms (2004) wrote in conclusion of his paper about Lake Corangamite:

> [...] the future for [the lake] is not without hope. It hinges on an appropriate decision [...] and then its speedy implementation. This will then allow [the lake] some latitude to vary as it always has done. Man will at last have learnt to live with a lake that fluctuates in extent and salinity.

We may hope that this optimistic statement could be extended to many other endangered water bodies all over the world, perhaps, not excluding the Aral Sea.

5.2   THE ARAL SEA AS AN OCEANOGRAPHIC OBJECT

Of course, the Aral Sea is not an ocean, nor even a "real" sea in the conventional sense of the word. However, we deliberately use the term oceanographic in the title of this section, as well as the entire book. On the one hand, the hydrophysical settings of the present Aral are provocative for oceanographic approaches to research. On the other hand, and reciprocally, new insight obtained from the Aral Sea can be quite instructive from the classical oceanography standpoint.

The most notable physical feature of the present Aral Sea is its strong stratification which develops, seasonally or at least intermittently, mainly because of the water exchanges between the eastern and western basins. This enhanced stratification greatly impedes vertical mixing and leads to hypoxic or anoxic conditions in the bottom part of the water column. The stability ratio $R_\rho = \beta \Delta S / \alpha \Delta T$ (where ΔT and ΔS are the temperature and salinity changes, α and β are the thermal expansion and salinity contraction coefficients) across the November 2002 pycnocline is about 7. Similar conditions (in terms of R_ρ) have been reported for the halocline in the central Baltic Sea (e.g., Lozovatsky, 1977), and for hot brines in the deep Red Sea, where diffusive stepwise structures have been observed.

Vertical gradients in salinity up to 12 ppt per 20 m have been documented for the Aral Sea, often accompanied by a temperature inversion up to 5°C (see Chapter 3). Such conditions are generally highly unusual for open seas and oceans, but they do

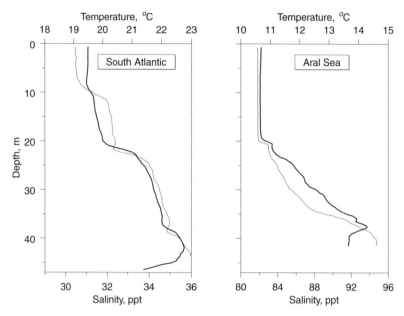

Figure 5.2. Temperature (bold curves) and salinity (light curves) profiles observed in the Aral Sea and in the coastal South Atlantic. (*left panel*) Southern Brazilian shelf north of the Plata estuary ("Rio Grande Current" area), 32°04′S, 51°00′W, 26 May, 2002, from Zavialov et al. (2003a). (*right panel*) Aral Sea (western trench), 45°05′N, 58°23′E, 12 November, 2002, from Zavialov et al. (2003b).

occur at specific locations, namely, in the areas adjacent to estuaries and river mouths, and possibly also in frontal zones of encounters between distinct water masses, in particular, near straits. For instance, the vertical profiles of temperature and salinity typical for the western Large Aral are remarkably reminiscent of those commonly observed in the coastal South Atlantic, on the shelf of southern Brazil near one of the world's largest estuaries. The Plata estuary drains the waters of Parana and Uruguay Rivers at the average rate of about 20,000 m^3/s (about 600 km^3/year), with peak values up to 100,000 m^3/s. Debouched from the estuary, this huge volume of fresh water then veers left under the action of the Coriolis force and propagates northward on the broad shelf as a highly stratified coastal current, sometimes referred to as the Rio Grande Current (Zavialov et al., 2002a). Exemplary profiles from the region are plotted together with those from the Aral Sea in Figure 5.2 on the same temperature and depth scale. Similarity between the profiles from Aral and from the ocean is striking. In both cases, the mechanism responsible for the vertical structure is connected with the reduction of vertical exchanges in the water column because of very strong haline stratification. In the Aral Sea case, the haline stratification is a consequence of excessive evaporation, especially in the eastern basin, and accumulation of salty water in the bottom layer of the western trench. In the South Atlantic shelf case, the haline stratification is due to abundant continental discharges from the river mouth accumulated in the near-surface layer. We see that, somewhat paradoxically, either deficit or excess of freshwater discharges both lead to a similar dynamical situation whose typical consequences, seen both in the Aral Sea and continental discharge controlled areas of the ocean, include enhanced seasonal and diurnal cycling, temperature inversions associated with the autumn cooling, and effective isolation of subsurface layers from atmospheric forcing. The latter factor is manifested, in particular, through relatively low coherence between the local wind stress and subsurface currents, which is typical for the present Aral Sea (Chapter 3) and the oceanic regions adjacent to large estuaries (e.g., Zavialov et al., 2002a). In such a situation, the upper layer acts much as a "rigid lid".

Kindred effects have been reported for the region near the mouth of the Mississippi River, for example. With its drainage basin of 2.98 million km^2, Mississippi is the largest river in North America. The freshwater discharge from the delta is debouched into the Gulf of Mexico and then flows westward along the Louisiana coast as the Louisiana Coastal Current (e.g., Wiseman et al., 2004). The current is strongly stratified: vertical changes of σ_t as large as up to $10 \, kg/m^3$ per 13 m have been reported, which is comparable with that in the Aral Sea. The current measurements suggest surface trapping of the energy, especially at high frequencies, as reported by Wiseman et al. (2004), who also emphasized the significance of shear-induced turbulent mixing in the bulk of the water column. This also fully applies to the conditions of the Aral Sea where the wind-generated shear currents and associated turbulence are believed to be major processes that govern vertical mixing. Lateral mixing, interleaving and layering taking place in the frontal zone between the waters of the western basin and the hyperhaline inflow from the eastern basin significantly affect the mesoscale dynamics. It has been suggested (Lilover et al., 1993) that lateral boundaries of mesoscale structures such as eddies, lenses, jets, and intrusions constitute "mixing vents" (i.e., narrow zones of intense mixing). To this end, mixing scenarios taking place under highly stratified, extreme conditions of the Aral Sea are of general interest.

Hypoxic or anoxic conditions develop each summer below the pycnocline on the Louisiana shelf (Wiseman et al., 1997), because the ventilation from the surface is greatly reduced due to the massive buoyancy flux delivered from the river. Plankton blooms in summer, and the products of their vital functions sink to the bottom and their oxidization consumes the dissolved oxygen which cannot be replenished because the lower layers are effectively isolated from the surface. The hypoxic conditions in the region and related processes have been observed for decades (e.g., Nelsen et al., 1994), but there are strong indications that the magnitude and spatial extent of the phenomenon are increasing, possibly in response to growing anthropogenic impacts in the river basin. Indeed, the area of the hypoxic water has doubled since 1985 and now exceeds 20,000 km^2. Some data point towards significant alterations in phytoplankton communities and even fisheries in the region (Rabalais et al., 2002; Wiseman et al., 2004).

Similar processes have been observed near other major estuaries. One of the most striking examples is the north-western shelf of the Black Sea, between the

Danube and Dnepr deltas, where hypoxia is believed to be the cause for mortality of up to 200 tonnes of biota (including fishes) per km^2 annually (e.g., Faschuk, 1995). The density stratification in the region in summer sometimes exceeds $2\,kg/m^3$ per 1 m (Selin et al., 1988), with the pycnocline located at a depth of 4–12 m. Sulphate-reduction of the organic matter under the hypoxic conditions leads to hydrogen sulphide contamination of the bottom layer, where the H_2S concentrations are up to $2\,ml/l$ (Faschuk, 1995). We note that the deep portion of the Black Sea (where saltier Mediterranean water inflowing through the Bosporus Strait, in a sense, acts much like the eastern basin water in the western part of the Aral Sea) is the world's largest anoxic basin contaminated with H_2S. However, the hydrogen sulphide zone on the shelf has been shown to be local (i.e., associated with the local stratification resulting from the river discharges, rather than the penetration of waters containing H_2S from the deep part of the sea onto the shelf). According to Faschuk (1995), the necessary condition for anoxia in the region is that the Danube and Dnepr discharges in May through July exceed the respective values of 70 and $4\,km^3$. The spatial extent of the anoxic zone and typical H_2S concentrations have been progressively increasing over the last decades. The H_2S zone on the shelf forms by late summer, frequently since late 1960s and annually since 1978. This is reminiscent of the Aral Sea's H_2S zone discussed in Chapter 3. The H_2S concentrations characteristic of the Aral Sea, however, are higher by at least one order of magnitude. The biological, chemical, and physical mechanisms responsible for such a difference in the rates of H_2S build up in the two regions are open to discussion. The typical total content of hydrogen sulphide on the shelf near the Dnepr mouth is estimated to be about 15,000–60,000 tonnes (Faschuk et al., 1986) (cf. Chapter 3 where we obtained 500,000 tonnes as the respective estimate for the typical H_2S content in the western basin of the Large Aral).

Anoxic conditions kindred to those of the Aral Sea can also be seen in a number of other strongly stratified oceanic regions. Another particular example that is worth mentioning are some fjords. In a typical fjord, the water column consists of a few meters thick brackish upper layer which receives freshwater runoff from rivers and streams, and a quasi-homogeneous layer of saltier water below it. At the entrance of the fjord, a sill is commonly found impeding the deep water circulation and exchanges with the outer region. As a consequence, many fjords exhibit stagnation and either seasonal or permanent oxygen depletion in the bottom layer (Colmen and Cushman-Roisin, 1999). Typical vertical gradients of salinity and density in the fjords are about 1 ppt and $1\,kg/m^3$ per 1 m, which is very much comparable with the respective parameters for the Aral Sea. As often seen in the Aral Sea, in autumn, the pycnocline in the fjords is commonly accompanied by a temperature inversion of 4–8°C (Colmen and Cushman-Roisin, 1992).

In conclusion of this chapter, we emphasize again that the Aral Sea desiccation and related dynamical phenomena should be thought of as reflections of processes manifested at a larger scale in the global perspective. The present Aral is therefore a good metaphor for studying hydrophysical and hydrochemical processes in "extreme" marine and lacustrine environments, affected by severe anthropogenic or climatic pressures.

6

Concluding remarks

The Aral Sea has lost about 3/4 of its area and nearly 9/10 of its volume since 1961. Simultaneously, the salinity of the lake's waters has increased by about an order of magnitude (except in the Small Sea, the northernmost portion of the lake that detached from the main part in 1989), which ranks present Aral among the saltiest large inland water bodies on earth. Although the catastrophic shallowing was triggered by a local forcing, both anthropogenic and climatic, the desiccation can be viewed as another striking manifestation of worldwide trends of global change, given that a large number of other lakes and even marine and oceanic regions have experienced environmental challenges of a kindred nature in the 20th century.

The morphological changes and salinization have led to a drastic reorganization of many physical processes taking place in the Aral's brines, from circulation patterns to mixing scenarios, and from thermohaline regime to land–lake–atmosphere exchanges. From the physical oceanography viewpoint, the present Aral Sea is a peculiar, complex and underexplored object having little in common with "original" Aral whose pre-desiccation physical regime has been thoroughly described and was well understood at the time. Obvious importance of many practical and scientific issues related to the Aral Sea crisis implies the need for comprehensive, updated information about the lake, which can be obtained through collection and analysis of new hydrological, meteorological, chemical, and biological data.

The interdisciplinary and international research of the last few years, encompassing field campaigns as well as remote sensing and modeling studies, have yielded significant insight into the physical and chemical states of the present Aral Sea, revealing its "new" thermal regime and thermohaline structure, circulation patterns, intense water and salt exchanges between the separate basins, periodically arising anoxic conditions and H_2S contamination in the bottom layer, and other interesting features, some of which were unexpected.

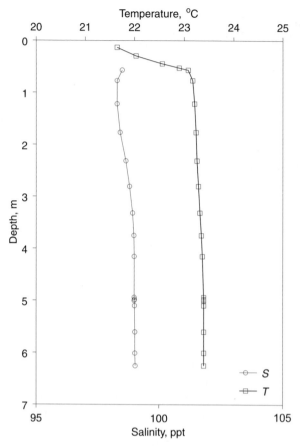

Figure 6.1. Vertical profiles of temperature (bold curve, boxes) and salinity (fine curve, circles) in the deepest portion of the strait connecting the eastern and western basins of the Large Aral Sea on 7 August, 2004.

As an illustration of the "last minute" findings, we add here some figures showing the most recent data acquired in the strait connecting the eastern and western basins. The data are still being processed at the time of writing. One of the most exciting results was the discovery of a narrow (a few hundred meters wide), deep (up to 8 m!) channel on the bed of the otherwise shallow (~1.5 m), flat bottom strait. Such a feature is amazing, considering that, as discussed in Section 3.1, the existing bathymetry maps had suggested that the depth of the strait should have been 2 m at most.

The vertical profiles of temperature and salinity obtained from the measurements in the deep channel are shown in Figure 6.1. The water column is well mixed throughout, except the uppermost 0.5 m where a temperature inversion as large as 1.5°C can be seen. The inversion may either be a result of meteorologically

Figure 6.2. Bottom relief section across the strait connecting the eastern and western basins of the Large Aral Sea, from the Kulandy Peninsula. Note the deep channel in the central part of otherwise shallow strait.

Based on seismic profiling data by Danis Nourgaliev (Kazan State University) (pers. commun.), and echo-soundings by the author.

forced surface cooling, or, otherwise, have an advective nature giving a hint of the colder western basin water propagating east along the strait as a surface current on top of warmer and saltier water.

The most plausible (if not the only possible) cause for the emergence of the narrow deep channel in the strait is the erosion of the lake floor by intense currents during the course of the interbasin water exchanges. Seismic profiling undertaken in August 2004 by a group from the Kazan State University led by D. Nourgaliev revealed the presence of the deep channel not only in the strait (Figure 6.2) but also on the bottom of the western basin itself, south of the Chernyshov Bay (not shown). This is a striking visual trace proving the existence of intense gravity currents drawing dense, salty eastern basin waters downslope into the western trench, as previously suggested by TS analyses (Chapter 3).

The observed erosive deepening of the strait connecting the two basins is of great importance, given that the expectations for the foreseen complete separation of the basins may need to be revisited in this context. Consequently, if the water and salt exchanges between the basins remain significant, the general prognostic considerations addressed in Chapter 4 without taking the channel deepening into account may need to be tuned accordingly. Therefore, understanding the mechanisms and dynamics of the interbasin exchanges (including the fluxes between the two parts of Large Aral and also occasional spills of Small Aral water into the Large Aral Sea) and related bed erosion processes, is among the foremost tasks.

Other priority tasks of the forthcoming research encompass, but are not limited to, quantification of the feedbacks discussed in Chapter 4, including substantiated assessment of the lake–groundwater exchanges, further specification of the lake circulation and thermohaline structure (especially in the eastern basin which is still largely a "white spot"), and comprehension of the chemical metamorphization processes accompanying the progressive salinization. An important general issue to be addressed is the correct coupling of physical, chemical, and biological processes taking place in the crisis-ridden Sea, in particular, those controlling the variability of Aral's anoxic zone.

New important and intriguing findings undoubtedly await the Aral Sea researchers in the near future. These will have important implications for similar studies worldwide.

References

Akhmedsafin U. M., Sadykov Zh. S., and Polomoshnova M. V. (1983) *Underground Water and Salt Exchanges in Aral Sea Basin: Present State and Forecast. Alma-Ata, Nauka,* 159 pp [in Russian].

Al Weshah R. A. (2000) The water balance of the Dead Sea: An integrated approach. *Hydrological Processes,* **14**, 145–154.

Aladin N. V. and Plotnikov I. S. (1995a) On the problem of possible conservation and rehabilitation of the Small Aral Sea. *Proceedings of the Zoological Institute, St. Petersburg,* **262**, 3–16 [in Russian].

Aladin N. V. and Plotnikov I. S. (1995b) Changes in Aral Sea level: paleolymnological and archeological evidences. *Proceedings of the Zoological Institute, St. Petersburg,* **262**, 17–46 [in Russian].

Aladin N. V., Plotnikov I. S., and Letolle R. (2004) Hydrobiology of the Aral Sea. In: J. C. J. Nihoul, P. O. Zavialov, and P. P. Micklin (eds), *Dying and Dead Seas.* NATO ARW/ASI Series, Kluwer Publishing, Dordrecht, pp. 125–158.

Anati D. A. (1997) The hydrography of a hypersaline lake. In: T. Niemi, Z. Ben-Avraham, and G. R. Gat (eds), *The Dead Sea—The Lake and its Settings.* Oxford University Press, Oxford, UK, pp. 89–103.

Anati D. A. (1998) Dead Sea water trajectories in the T–S space. *Hydrobiologia,* **381**, 43–49.

Asarin A. E. (1973) Components of Aral Sea's water balance and their influence on inter-annual variability of the sea level. *Vodniye Resursy,* **5**, 29–40 [in Russian].

Ashirbekov U. and Zonn I. (2003) *Aral: The History of Vanishing Sea.* Edel-M, Moscow-Nukus, 63 pp [in Russian].

Barth A. (2000) Modélisation Mathématique et Numérique de la Mer d'Aral. M.Sc. Thesis. University of Liège, Belgium, 138 pp. [in French].

Barthold V. V. (1902) Information about the Aral Sea and low reaches of Amu-Darya from the ancient times to XVII century. *Izvestiya Turkestanskogo Otdeleniya Russkogo Geograficheskogo Obschestva (Proceedings of Turkestan Branch of Russian Geographic Society),* **4**(2), 1–120 [in Russian].

Baypakov K. M., Boroffka N., Savelieva T. V., Akhatov G. A., Lobas D. A., Erzhanova A. A. (2004) Results of archeological studies in the northern Priaralye Region under the INTAS

"Climan" Project. *Izvestiya, National Academy of Sciences of Kazakhstan*, **242**(1), 236–254 [in Russian].

Benduhn F. and Renard F. (2004) A dynamic model of the Aral Sea water and salt balance. *Journal of Marine Systems*, **47**(1–4), 35–50.

Berg L. S. (1908) *Aral Sea. An Attempt of Physical Geographic Description*. Izvestiya Turkestanskogo Otdeleniya, Russian Geographic Society, St. Petersburg, **5**, 9, 580 pp [in Russian].

Betger E. K. (1953) *Diaries of A. I. Butakov Expedition on "Konstantin" Schooner for Prospecting the Aral Sea in 1848–1849*. Tashkent, 53 pp [in Russian].

Beyth M. (1980) Recent evolution and present stage of the Dead Sea brines. In: A. Nissenbaum (ed.), *Hypersaline Brine and Evaporitic Environments*. Elsevier, Amsterdam, pp. 155–166.

Björklund G. (1999) The Aral Sea—water resources, use and misuse. In: K. Lindhal Kiessling (ed.), *Alleviating the Consequences of an Ecological Catastrophe*. Swedish Unifem–committee, Stockholm, pp. 42–50.

Blinov L. K. (1956) *Hydrochemistry of Aral Sea*. Gidrometeoizdat, Leningrad, 152 pp [in Russian].

Bogomolets N. A. (1903) On bacterial flora of Aral Sea. In: *Scientific Results of the Aral Expedition* (vol. 4). St. Petersburg [in Russian].

Boomer I., Aladin N. V., Plotnikov I., and Whatley R. (2001) The paleolimnology of the Aral Sea: A review. *Quarterly Science Review*, **19**, 1259–1278.

Bortnik V. N. (1996) Changes in water level and hydrological balance of the Aral Sea. In: P. P. Micklin and W. D. Williams (eds), *The Aral Sea Basin*. NATO ASI Series, Partnership Sub-Series, 2. Environment, **12**, Springer Verlag, Berlin, pp. 25–32.

Bortnik V. N. and Dauletiyarov K. Zh. (1985) *Numerical Study of the Aral Sea Circulation*. Vycheslitsel'niy Tsentr AN, Moscow, 36 pp [in Russian].

Bortnik V. N. and Chistyaeva S. P. (eds) (1990) *Hydrometeorology and Hydrochemistry of the USSR Seas. Vol. VII: The Aral Sea*. Gidrometeoizdat, Leningrad, 196 pp. [in Russian].

Briegleb B. P. (1992) Delta-Eddington approximation for solar radiation in the NCAR Community Climate Model. *Journal of Geophysical Research*, **97**(D7), 7603–7612.

Butakov A. I. (1853) Information on the expedition organized for assessing the Aral Sea in 1848. *Vestnik Russkogo Geograficheskogo Obschestva*, Part 7, Book 1, Section 7, pp. 1–9 [in Russian].

Calder I. R. and Neal C. (1984) Evaporation from saline lakes: a combination equation approach. *Hydrological Sciences*, **29**, 89–97.

Carmouze J. P., Durand J. R., and Lévêque C. (1983) *Lake Chad*. Junk, The Hague, 575 pp.

Central Asian States (2000) State of Environment in the Aral Sea Basin. Regional report of the Central Asian States. http://www.grida.no/aral (February 2004).

Chernenko I. M. (1970) Inflow of groundwater into the Aral Sea and its significance in mitigating the Aral Sea crisis. *Problemy Osvoeniya Pustyn'*, **4**, 28–38 [in Russian].

Chernenko I. M. (1972) On groundwater inflow, salt balance, and Aral crisis. *Problemy Osvoeniya Pustyn'*, **2**, 32–42 [in Russian].

Chernenko I. M. (1983) Water–salt balance and the use of the desiccating Aral. *Problemy Osvoeniya Pustyn'*, **3**, 18–24 [in Russian].

Chub V. E. (2000a) *Climate Change and its Impact on Natural Resources Potential of Republic of Uzbekistan*, SANIGMI, Tashkent, 252 pp [in Russian].

Chub V. E. (2000b) Characteristics of the water resources in the Central Asia discharge formation zone and their variability. In: U. Umarov and A. Kh. Karimov (eds), *Water Resources, Aral Problem and Environment*. Universitet, Tashkent, pp. 3–18 [in Russian].

Coe M. T. and Foley J. A. (2001) Human and natural impacts on the water resources of the Lake Chad basin. *Journal of Geophysical Research*, **106**, 3349–3356.

Coleman and Cushman-Roisin (1992) A self-sustained pump across temperature–salinity gradients in coastal waters. *Ocean Engineering*, **19**(1), 57–74.

Coleman and Cushman-Roisin (1999) Energy from seawater: A self-sustained diffusive pump. *IOA Newsletter*, **10**(3), 1–7.

Darcy H. (1856) *Fontaines Publiques de la Ville de Dijon*. Victor Dalmont, Paris [in French].

Drumeva L. B. and Tsitsarin A. G. (1984) Present salt composition of the Aral and Azov seas. *Soviet Meteorology and Hydrology*, **3**, 112–115 [in Russian].

Faschuk D. Ya. (1995) Hydrogen sulphide zone on the northwestern shelf of the Black Sea: Nature, origin, and dynamical mechanisms. *Vodniye Resursy*, **22**(5), 568–584 [in Russian].

Faschuk D. Ya., Briantsev V. A., and Trotsenko B. G. (1986) *Anthropogenic Impacts on Coastal Marine Ecosystems*. VNIIRO, Moscow, 34 pp [in Russian].

Fedorov P. V. (1980) On some problems of the Holocene history of Caspian Sea and Aral Sea. In: *Moisture Variability in Aral–Caspian Region in Holocene*. Nauka, Moscow, pp. 19–22 [in Russian].

Filippov Yu. G. (1970) A method to calculate marine currents. *Trudy GOIN*, **103**, 87–94 [in Russian].

Fortunatov M. A. and Sergiyenko V. D. (1950) New data on morphometry of Aral Sea. *Izvestiya Vsesoyuznogo Geograficheskogo Obschestva*, **82**(1), 51–58 [in Russian].

Friedrich J. and Oberhänsli H. (2004) Hydrochemical properties of the Aral Sea water in summer 2002. *Journal of Marine Systems*, **47**(1–4), 77–88.

Froebrich J. and Kayumov O. (2004) Water management aspects of Amu-Darya. In: J. C. J. Nihoul, P. O. Zavialov, and P. P. Micklin (eds), *Dying and Dead Seas*. NATO ARW/ASI Series, Kluwer Publishing, Dordrecht, pp. 49–76.

Gavrieli I. (1997) Halite deposition in the Dead Sea: 1960–1993. In: T. Niemi, Z. Ben-Avraham, G. R. Gat (eds), *The Dead Sea—The Lake and its Settings*. Oxford University Press, Oxford, UK, pp. 161–170.

Gavrieli I., Starinsky A., and Bein A. (1989) The solubility of halite as a function of temperature in the highly saline Dead Sea brine system. *Limnology and Oceanography*, **34**, 1224–1234.

Gavrieli I. and Oren A. (2004) The Dead Sea as a dying lake. In: J. C. J. Nihoul, P. O. Zavialov, and P. P. Micklin (eds), *Dying and Dead Seas*. NATO ARW/ASI Series, Kluwer Publishing, Dordrecht, pp. 287–305.

Gertman I. and Hecht A. (2002) The Dead Sea hydrography from 1992 to 2000. *Journal of Marine Systems*, **35**, 169–181.

Gill A. E. (1982) *Atmosphere–Ocean Dynamics*. Academic Press, London, 662 pp.

Ginzburg A. I., Kostianoy A. G., and Sheremet N. A. (2003) Thermal regime of the Aral Sea in the modern period (1982–2000) as revealed by satellite data. *Journal of Marine Systems*, **43**, 19–30.

Giorgi F. and Shields C. (1999) Tests of precipitation parameterizations available in the latest version of the NCAR regional climate model (RegCM) over continental United States. *Journal of Geophysical Research*, **104**, 6353–6375.

Giorgi E., Marinucci M. R., and Bates G. T. (1993a) Development of a second generation regional climate model (RegCM2). Part I: Boundary layer and radiative transfer processes. *Monthly Weather Review*, **121**, 2794–2813.

Giorgi E., Marinucci M. R., and Bates G. T. (1993b) Development of a second generation regional climate model (RegCM2). Part II: Convective processes and assimilation of lateral boundary conditions. *Monthly Weather Review*, **121**, 2814–2832.

Giralt S., Ramon J., Klerkx J., Riera S., Leroy S., Buchaca T., et al. (2004) 1000-year environmental history of Lake Issyk–Kul. In: J. C. J. Nihoul, P. O. Zavialov, and P. P. Micklin (eds), *Dying and Dead Seas*. NATO ARW/ASI Series, Kluwer Publishing, Dordrecht, pp. 253–286.

Glazovskiy N. F. (1976) Ground water and ion discharge into Aral, Caspian, and Black Seas. *Doklady AN SSSR*, **227**, 4, 961–964 [in Russian].

Golmen L. G. and Cushman-Roisin B. (1992) A self-sustained pump across temperature–salinity gradients in coastal waters. *Ocean Engineering*, **19**(1), 57–74.

Golmen L. G. and Cushman-Roisin B. (1992) Energy from seawater: A self-sustained diffusive pump. *IOA Newsletter*, **10**(3), 1–7.

Golubtsov V. V. and Morozova O. A. (1972) On the present water budget of Aral Sea. *Trudy KazNIGMI (Kazakhstan Institute of Hydrometeorology)*, **44**, 87–99 [in Russian].

Goptarev N. P. and Panin G. N. (1970) Influence of temperature stratification in the near-water layer of the atmosphere on evaporation rates. *Trudy GOIN*, **98**, 148–155 [in Russian].

Gorodetskaya M. E. (1978) On the terraces of the Aral Sea. *Geomorfologiya*, **1**, 46–55 [in Russian].

Grigoriev A. A. and Lipatov V. B. (1982) Dynamics and sources of dust storms in Aral region according to observations from space. *Izvestiya AN SSSR*, Geographic Series, **5**, 93–98 [in Russian].

Hannan T. and O'Hara S. L. (1998) Managing Turkmenistan's Kara Kum canal: Problems and prospects. *Post-Soviet Geography and Economics*, **39**(4), 225–235.

Henderson–Sellers B. (1985) New formulation of eddy diffusion thermocline models. *Applied Mathematical Modelling*, **9**, 441–446.

Hostetler S. W. and Bartlein P. J. (1990) Simulation of lake evaporation with application to modeling lake level variations of Harney–Malheur Lake, Oregon. *Water Resources Research*, **26**, 2603–2612.

Hostetler S. W., Giorgi F., Bates G. T., and Bartlein P. J. (1994) Lake–atmosphere feedbacks associated with Paleolakes Bonneville and Lahontan. *Science*, **263**, 665–668.

Jarsjö J. and Destouni D. (2004) Groundwater discharge into the Aral Sea after 1960. *Journal of Marine Systems*, **47**(1–4), 109–120.

Inogamov Kh., Fuzaylov I. A., and Avazov Sh. (1979) Structure of the Earth crust in Eastern Ustyurt. *Uzbekskiy Geologicheskiy Zhurnal*, **2**, 46–56 [in Russian].

Ivanov V. A., Lyubartseva S. P., Mikhailova E. N., Shapiro N. B., and Gertman I. (2002) A model of the Dead Sea. Simulation of the thermohaline structure variability in 1992–2000. *Marine Hydrophysical Journal*, **5**, 3–23 [in Russian].

Kalnay E., Kanamitsu M., Kistler R., Collins W., Deaven D., Gandin L., et al. (1996) The NCEP/NCAR 40-year reanalysis project. *Bulletin of American Meteorological Society*, **77**, 437–472.

Khan V. M., Vilfand R. M., and Zavialov P. O. (2004) Long-term variability of air temperature in the Aral Sea region. *Journal of Marine Systems*, **47**(1–4), 25–34.

Khomerini I. V. (1969) The Monte Carlo method applied to investigating the regime of the Aral Sea. *Proceedings of 1st All-Union Conference on Monte Carlo Methods*, Metsnierba, Tbilisi, pp. 229–230 [in Russian].

Khomerini I. V. (1978) Aral Sea water budget modelling: The multi-dimensional case. *Meteorologiya i Gidrologiya*, **5**, 70–80 [in Russian].

Kira T. (1997) Survey of the State of World Lakes. In: S. E. Jorgensen and I. Matsui (eds), *Guidelines of Lake Management. Vol. 8. The World's lakes in Crisis*. ILEC Foundation/UNEP, pp. 147–155.

Kiselev V. P. (2004) Aral sea coastal region. GIS Laboratory, Nuclear Safety Institute. http://ibraent.ibrae.ac.ru/aral (May 2004).

Korganov A. S. (1969) Water balance of the lower reaches of Syr-Darya river, from Chardary to the Aral Sea. *Problemy Osvoeniya Pustyn'*, **5**, 56–64 [in Russian].

Kosarev A. N. (1975) *Hydrology of Caspian Sea and Aral Sea.* Moscow State University, Moscow, 272 pp. [in Russian].

Kosarev A. N. and Tsvetsinskiy A. S. (1976) Winter vertical circulation in the Aral Sea. In: *Complex Investigations of the Caspian Sea.* Moscow University, Moscow, pp. 212–219 [in Russian].

Kosarev A. N. and Yablonskaya E. A. (1994) *The Caspian Sea.* Academic Publishing, Hague, 259 pp.

Kostianoy A. G., Zavialov P. O., and Lebedev S. (2004) What do we know about dead, dying and endangered seas and lakes? In: J. C. J. Nihoul, P. O. Zavialov, and P. P. Micklin (eds), *Dying and Dead Seas.* NATO ARW/ASI Series, Kluwer Publishing, Dordrecht, pp. 1–48.

Kouraev A. V., Papa F., Mognard N. M., Buharizin P., Cazenave A., Cretaux J.-F., et al. (2004) Sea ice cover in the Caspian and Aral seas from historical and satellite data. *Journal of Marine Systems*, **47**(1–4), 89–100.

Krumgaltz B. S., Hecht A., Starinsky A., and Katz A. (2000) Thermodynamic constraints on Dead Sea evaporation: Can the Dead Sea dry up? *Chemical Geology*, **165**, 1–11.

Kvasov D. D. (1991) Basic information about the Aral Sea. In: V. A. Rumiantsev et al. (eds), *History of Lakes Sevan, Issyk-Kul, Balkhash, Zaisan, and Aral.* Nauka, Leningrad, pp. 239–243 [in Russian].

Kvasov D. D. and Mamedov E. D. (1991) Progress of irrigation and its influence on the water balance. In: V. A. Rumiantsev (eds), *History of Lakes Sevan, Issyk-Kul, Balkhash, Zaisan, and Aral.* Nauka, Leningrad, pp. 227–230 [in Russian].

Lemoalle J. (1991) The hydrology of Lake Chad during a drought period (1973–1989). *FAO Fisheries Reports*, **445**, 54–61.

Lemoalle J. (2004) Lake Chad: A changing environment. In: J. C. J. Nihoul, P. O. Zavialov, and P. P. Micklin (eds), *Dying and Dead Seas.* NATO ARW/ASI Series, Kluwer Publishing, Dordrecht, pp. 321–339.

Létolle R. and Mainguet M. (1993) *Aral.* Springer Verlag, Paris, 357 pp.

Létolle R. and Mainguet M. (1996) *Der Aralsee—Eine Ökologische Katastrophe.* Springer Verlag, Berlin [in German].

L'Hôte Y., Mahé G., Somé B., and Triboulet J.-P. (2002) Analysis of a Sahelian annual rainfall index from 1896 to 2000: The drought continues. *Hydrological Sciences Journal*, **47**, 563–572.

Lilover M., Lozovatsky I. D., Gibson C. H., and Nabatov V. N. (1993) Turbulent exchange through the Equatorial Undercurrent core of the Central Pacific. *Journal of Marine Systems*, **4**, 183–195.

Lozovatsky I. D. (1977) Investigation of small-scale temperature inhomogeneities in the southern part of the Baltic Sea. *Oceanology* (English edition), **17**(2), 135–138.

Lvov V. P. (1959) Variability of Aral Sea level over the last 100 years. *Trudy GOIN*, **46**, 80–114 [in Russian].

Lvov V. P. (1966) Statistical characterization for a long record of annual mean level of the Aral Sea. *Oceanology*, **6**(4), 632–640 [in Russian].

Lvov V. P. (1970a) Seasonal variability of the Aral Sea level. *Trudy GOIN*, **98**, 107–116 [in Russian].

Lvov V. P. (1970b) On relation between ion (salt) compositions of the Aral Sea water and ground water of the Priaralye region (On the the groundwater component of the water and salt budgets of the Aral Sea). *Trudy GOIN*, **101**, 87–100 [in Russian].

Lvov V. P., Krylova Z. A., and Smirnova R. V. (1970) Water budget of the Aral Sea (monthly, annual, and long-term values for 1952–1966). *Trudy GOIN*, **101**, 5–33 [in Russian].

Lvovich M. I. and Tsigelnaya I. D. (1978) Controlling the water balance of the Aral Sea. *Izvestiya AN SSSR*, Geography series, **1**, 42–54 [in Russian].

Lymarev V. I. (1967) *Shores of the Aral Sea, an Inland Arid Zone Water Body*. Leningrad, Nauka, 252 pp. [in Russian].

Mamedov E. and Trofimov G. N. (1986) On the problem of long-term oscillations of the Middle Asia river discharges. *Problemy Osvoeniya Pustyn'*, **1**, 12–16 [in Russian].

Mellor G. L. (1992) Users's guide for a three-dimensional, primitive equation, numerical ocean model. In: *Progress in Atmospheric and Oceanic Sciences*, Princeton University Press, Princeton, NJ, 35 pp.

McClain E. P. (1989) Global sea surface temperature and cloud clearing for aerosol optical depth estimates. *International Journal of Remote Sensing*, **10**, 767–769.

Micklin P. P. (2004) The Aral Sea crisis. In: J. C. J. Nihoul, P. O. Zavialov, and P. P. Micklin (eds), *Dying and Dead Seas*. NATO ARW/ASI Series, Kluwer Publishing, Dordrecht, pp. 99–123.

Mikhailov V. N., Kravtsova V. I., Gurov F. N., Markov D. V., and Gregoire M. (2001) Assessment of the present-day state of the Aral Sea. *Vestnik Moskovskogo Universiteta*, Geographic Series, **6**, 14–21 [in Russian].

Mirabdullaev I. M., Joldasova I. M., Mustafaeva Z. A., Kazakhbaev S., Lyubimova S. A., and Tashmukhamedov B. A. (2004) Succession of the ecosystems of the Aral Sea during its transition from oligohaline to polyhaline waterbody. *Journal of Marine Systems*, **47**(1–4), 101–108.

Muminov F. A. and Inagatova I. I. (1995) *Variability of Central Asian Climate*, SANIGMI, Tashkent, 215 pp [in Russian].

Nelsen T. A., Blackwelder P., Hood T., McKee B., Romer N., Alvarez–Zarikian C., and Metz S. (1994) Time-based correlation of biogenic, lithogenic, and authigenic sediment components with anthropogenic inputs in the Gulf of Mexico NECOP study area. *Estuaries*, **17**(4), 873–885.

Neev D. and Emery K. O. (1967) *The Dead Sea Depositional Processes and Environments of Evaporites*. Geological Survey of Israel, Tel-Aviv, 147 pp.

Nihoul J. C. J., Kosarev A. N., Kostianoy A. G., and Zonn I. S. (eds) (2002) *The Aral Sea: Selected Bibliography*. Noosphere, Moscow, 232 pp.

Nihoul J. C. J., Zavialov P. O., and Micklin P. P. (eds) (2004) *Dying and Dead Seas*. NATO ARW/ASI Series, Kluwer Publishing, Dordrecht, 384 pp.

Nikolskiy G. V. (1940) *Fishes of Aral Sea*. MOIP, Moscow, 216 pp.

Nurtaev B. (2004) Aral Sea basin evolution: Geodynamical aspect. In: J. C. J. Nihoul, P. O. Zavialov, and P. P. Micklin (eds), *Dying and Dead Seas*. NATO ARW/ASI Series, Kluwer Publishing, Dordrecht, pp. 91–97.

Oceanographic Tables (1975) Gidrometeoizdat, Leningrad, 477 pp. (edited by G. S. Ivanov).

Oceanological Tables for Caspian, Aral, and Azov Seas. (1964) Gidrometeoizdat, Moscow, 107 pp [in Russian].

Oren A. (1993) The Dead Sea—Alive again. *Experientia*, **49**, 518–522.

Oren A. (1999) Microbiological studies in the Dead Sea: Future challenges towards the understanding of life at the limit of salt concentrations. *Hydrobiologia*, **205**, 1–9.

Pala C. (2003) $85 Million project begins for revival of the Aral Sea. *New York Times*, August 4, p. F3.

Peneva E. L., Stanev E. V., Stanichniy S. V., Salokhiddinov A., and Stulina G. (2004) The recent evolution of the Aral Sea level and water properties: Analysis of satellite, gauge, and hydrometeorological data. *Journal of Marine Systems*, **47**(1–4), 11–24.

Petr T. (1992) Lake Balkhash, Kazakhstan. *International Journal of Salt Lake Research*, **1**(1), 21–46.

Pickard G. L. and Emery W. J. (1990) *Descriptive Physical Oceanography* (5th edition). Butterworth–Heinemann, Oxford, UK, 323 pp.

Pinkhasov B. I. (2000) Paleogeography of Turan and Tien Shan plains in Neogene. *Geologiya va Mineral Resurslar, Tashkent*, **6**, 3–8 [in Russian].

Pshenin G. N., Steklenkov A. P., and Cherkinskiy A. E. (1984) Origin and age of pre-Holocene terraces of the Aral Sea. *Doklady AN SSSR*, **276**, 3, 675–677 [in Russian].

Rabalais N. N., Turner R. E., and Scavia D. (2002) Beyond science into policy: Gulf of Mexico hypoxia and the Mississippi River. *Bioscience*, **52**, 129–142.

Rafikov A. A. and Tetyukhin G. F. (1981) *Lowering of the Aral Sea Level and Environmental Changes in the Low Amu-Darya Region*. FAN, Tashkent, 199 pp [in Russian].

Ressl R. and Micklin P. P. (2004) Morphological changes in the Aral Sea: Satellite imagery and water balance model. In: J. C. J. Nihoul, P. O. Zavialov, and P. P. Micklin (eds), *Dying and Dead Seas*. NATO ARW/ASI Series, Kluwer Publishing, Dordrecht, pp. 77–90.

Revina S. K., Bakum T. A., and Zaklinskiy A. B. (1970) Main features of salinity distribution of the Aral Sea inferred from data for 1952–1966. *Trudy GOIN*, **101**, 80–86 [in Russian].

Rodin D. A., Zavialov P. O., and Zhurbas V. N. (2005) Modeling meso- and basin-scale circulations in the present Aral Sea using Princeton Ocean Model. *Physical Ecology*, Moscow State University, accepted for publication, in press [in Russian].

Rogov M. M., Khodkin S. S., and Revina S. K. (1968) Hydrology of Amu-Darya Mouth Area. *Trudy GOIN*, **94**, 268 pp [in Russian].

Romanovsky V. V. (2002) Water level variations and water balance of lake Issyk–Kul. In: J. Klerkx and B. Imanackunov (eds), *Lake Issyk–Kul: Its Natural Environment*. NATO Science Series, Kluwer Academic Publishers, Netherlands, pp. 45–58.

Romashkin V. S. and Samoilenko V. S. (1953) Hydrometeorological characteristics of the Aral Sea. *Trudy GOIN*, **12**, 98–129 [in Russian].

Rubanov I. V. and Timokhina N. I. (1982) Conditions for formation of mirabilite (Aral Sea as an example). *Zapiski Uzbekistanskogo Otdeleniya VMO*, **35**, 57–60 [in Russian].

Rubanov I. V. and Bogdanova N. M. (1987) A quantitative estimate of salt deflation from the dried bottom of the Aral Sea. *Problemy Osvoeniya Pustyn'*, **3**, 9–16 [in Russian].

Rubanov I. V., Ishniyazov D. P., Baskakova M. A., and Chistyakov P. A. (1987) *Geology of Aral Sea*. FAN, Tashkent, 247 pp [in Russian].

Rubinova F. E. (1987) Influence of water meliorations on runoff and hydrochemical regime of Aral Sea basin rivers. *Trudy SANII*, **124**(205), 160 pp. [in Russian].

Sadov A. A. and Krasnikov V. V. (1987) Detection of foci of subaqueous groundwater discharge into Artal Sea by remote sensing methods. *Problemy Osvoeniya Pustyn'*, **1**, 28–36 [in Russian].

Sailing Directions for the Aral Sea (1963) Izdatelstvo Gidrograficheskoy Sluzhby VMF, Leningrad, 148 pp [in Russian].

Samoilenko V. S. (1953) Analysis of the Aral Sea's heat budget. *Trudy GOIN*, **12**, 98–129 [in Russian].

Samoilenko V. S. (1955a) On forthcoming changes in temperature regime of the Aral Sea. *Trudy GOIN*, **20**, 190–246 [in Russian].

Samoilenko V. S. (1955b) State-of-the-art of the water budget and level variability problems for the Aral Sea. *Trudy GOIN*, **20**, 127–166 [in Russian].

Sarch M. T. and Birkett C. (2000) Fishing and farming at Lake Chad: Responses to lake level fluctuations. *Geographic Journal*, **166**, 156–172.

Selin P. Yu., Ayzatullin T. A., Leonov A. V., and Faschuk D. Ya. (1988) Chemical dynamics of the hydrogen sulphide zone on the northwestern shelf of the Black Sea. *Vodniye Resursy*, **16**(4), 144–173 [in Russian].

Shaporenko S. I. (1995) Balkhash Lake. In: A. T. Mandych (ed.), *Enclosed Seas and Large Lakes of Eastern Europe and Middle Asia*. SPB Academic Publishing, Amsterdam, pp. 155–198.

Shkudova G. YA. and Kovalev N. P. (1969) An attempt of implementation of a stationary hydrodynamical model for computing currents in a shallow sea. *Meteorologiya i Gidrologiya*, **10**, 76–86 [in Russian].

Shults V. L. (1975) Water budget of Aral Sea. *Trudy SANIGMI*, Central Asian Institute of Hydrometeorology, Tashkent, **23**(104), 3–28 [in Russian].

Shults V. L. and Shalatova L. I. (1964) Water budget of Aral Sea. *Trudy TashGU*, Tashkent State University, **269**, 33–58 [in Russian].

Simonov A. I. (1954) On origins of anticyclonic circulation of the Aral Sea waters. *Meteorologiya i Gidrologiya*, **2**, 50–52 [in Russian].

Simonov A. I. (1962) The origin of relatively high-salinity water of the Aral Sea western depression. *Trudy GOIN*, **68**, 103–117 [in Russian].

Simonov A. I. and Goptarev N. P. (eds) (1972) Present and Prospective Water and Salt Budgets of Southern Seas of USSR. Gidrometeoizdat, Moscow, 236 pp. (*Trudy GOIN*, **108**) [in Russian].

Sirjacobs D., Gregoire M., Delhez E., and Nihoul J. C. J. (2004) Influence of the Aral Sea negative water balance on its seasonal circulation patterns: Use of a 3D hydrodynamic model. *Journal of Marine Systems*, **47**(1–4), 51–66.

Small E. E., Giorgi F., and Sloan L. C. (1999) Regional climate model simulations of precipitation in central Asia: Mean and interannual variability. *Journal of Geophysical Research*, **104**, 6563–6582.

Small E. E., Sloan L. C., Giorgi F., and Hostetler S. (1999) Simulating the water balance of the Aral Sea with a coupled regional climate lake model. *Journal of Geophysical Research*, **104**, 6583–6602.

Small E. E., Giorgi F., Sloan L. C., and Hostetler S. (2001a) The effects of desiccation and climatic change on the hydrology of the Aral Sea. *Journal of Climate*, **14**, 300–332.

Small E. E., Sloan L. C., and Nychka D. (2001b) Changes in surface air temperature caused by desiccation of the Aral Sea. *Journal of Climate*, **14**, 284–299.

Sopach E. D. (1958) *Electric Conductivity as a Means for Determining Salinity of Sea Waters*. Gidrometeoizdat, Moscow, 139 pp [in Russian].

Stanev E. V., Peneva E. L., and Mercier F. (2004) Temporal and spatial patterns of sea level in inland basins: Recent events in the Aral Sea. *Geophysical Research Letters*, accepted for publication, in press.

Steinhorn I. (1983) In situ salt precipitation at the Dead Sea. *Limnology and Oceanography*, **28**, 580–583.

Steinhorn I., Assaf G., Gat J. R., Nishri A., Nissenbaum A., Stiller M., et al. (1979) The Dead Sea: Deepening of the mixolimnion signifies the overture to overturn of the water column. *Science*, **206**, 55–57.

Steinhorn I. and Gat J. R. (1983) The Dead Sea. *Scientific American*, **249**, 102–109.

Talling J. F. and Lemoalle J. (1998) *Ecological Dynamics of Tropical Inland Waters.* Cambridge University Press, Cambridge, UK, 441 pp.

Tilho J. (1928) Variations et disparition possible du lac Tchad. *Annales Géographie*, **37**, 238–260 [in French].

Tillo A. A. (1877) *Description of Aral–Caspian Levelling Performed in 1874 at Request of the Russian Geographic Society and its Orenburg Division.* Russian Geographic Society, St. Petersburg, 42 pp [in Russian].

Timms B. W. (2004) The continued degradation of Lake Corangamite, Australia. In: J. C. J. Nihoul, P. O. Zavialov, and P. P. Micklin (eds), *Dying and Dead Seas.* NATO ARW/ASI Series, Kluwer Publishing, Dordrecht, pp. 307–319.

Vaynbergs I. G. and Stelle V. Ya. (1980) Late Quaternary stages of the Aral Sea development and their connections with the changes of climate conditions at the time. In: *Moisture Variability in Aral–Caspian Region in the Holocene.* Nauka, Moscow, pp. 176–180 [in Russian].

Veselov V., Panichkin V., Truschel L., Zaharova N., Kalmykova N., Vinnikova T., and Miroshnichenko O. (2002) Simulation of groundwater resources of Aral Sea basin. UNESCO project 00KZ11102 report. Institute of Hydrogeology and Hydrophysics, Ministry of Education and Science, Kazakhstan. http://www.aralmodel.unesco.kz (May 2004).

Vinogradov M. E., Shushkina E. A., Vostokov S. V., Vereschaka L. A., and Lukasheva T. A. (2002) Population dynamics of the ctenophores Mnemiopsis leidyi and Beroe ovata near the Caucasus shore of the Black Sea. *Oceanology*, **42**(5), 693–701 [in Russian].

Volftsun I. B. and Sumarokova V. V. (1985) Dynamics of anthropogenic and natural losses of Amu-Darya and Syr-Darya runoffs over a long period. *Meteorologiya i Gydrologiya*, **2**, 98–104 [in Russian].

Volkov I. I., Skirta A. Yu., Makkaveev P. N., Demidova T. P., and Rozanov A. G. (2002) On physical and chemical homogeneity of the deep water in the Black Sea. In: A. G. Zatsepin and M. V. Flint (eds), *Multidisciplinary Investigations of the Northern Part of the Black Sea.* Nauka, Moscow, pp. 161–169 [in Russian].

Wang H. (1993) Deforestation and desiccation in China: A preliminary study. www.library.utoronto.ca/pcs/state/chinaeco/forest.htm (May 2004).

Wheeler S. S. (1974) *The Desert Lake: The Story of Nevada's Pyramid Lake.* Caxton Printers, Caldwell, Idaho, 150 pp.

Williams W. D. (1986) Conductivity and salinity of Australian salt lakes. *Australian Journal of Marine and Freshwater Research*, **37**, 177–182.

Williams W. D. (1995) Lake Corangamite, Australia, a permanent saline lake: Conservation and management issues. *Lakes and Resevoirs: Research and Management*, **1**, 55–64.

Wiseman W. J., Rabalais N. N., Turner R. E., Dinnel S. P., and MacNaughton A. (1997) Seasonal and interannual variability within the Lousiana Coastal Current: Stratification and hypoxia. *Journal of Marine System*, **12**, 237–248.

Wiseman W. J., Rabalais N. N., Turner R. E., and Justic D. (2004) Hypoxia and the physics of the Louisiana Coastal Current. In: J. C. J. Nihoul, P. O. Zavialov, and P. P. Micklin (eds), *Dying and Dead Seas.* NATO ARW/ASI Series, Kluwer Publishing, Dordrecht, pp. 359–372.

World Lakes Database (2004) www.ilec.or.jp/database/database.htm (May 2004).

Yanshin A. L. and Goldenberg L. A. (1963) *First Russian Scientific Studies of Ustyurt.* Izdatel'stvo AN SSSR, Moscow, 326 pp. [in Russian].

Yechieli Y., Gavrieli I., Berkowitz B., and Ronen D. (1998) Will the Dead Sea die? *Geology*, **26**, 755–758.

Zhitomirskaya O. M. (1964) *Climatic Description of Aral Sea Region*. Gidrometeoizdat, Leningrad, 67 pp [in Russian].

Zaikov B. D. (1946) Present and future water budget of Aral Sea. *Trudy NIU GUGMS*, Ser. 4, **39**, 25–59 [in Russian].

Zaikov B. D. (1952) Water budget and level of Aral Sea in the context of the construction of the Main Turkmenistan Channel. *Trudy GGI*, **16**, 44 pp. [in Russian].

Zavialov P. O., Möller Jr. O. O., and Campos E. (2002a) First direct measurement of currents on the continental shelf of Southern Brazil. *Continental Shelf Research*, **22**, 14, 1975–1986.

Zavialov P. O., Grigorieva J. V., Möller Jr. O. O., Kostianoy A. G., and Gregoire M. (2002b) Continuity preserving modified maximum cross-correlation technique. *Journal of Geophysical Research*, doi:10.1029/2001JC0011116, **107**(C10), 1–10.

Zavialov P. O., Kostianoy A. G., and Möller Jr. (2003a) SAFARI cruise: Mapping river discharge effects on Southern Brazilian shelf. *Geophysical Research Letters*, doi: 1029/2003GL018265, **30**(21), 1–4.

Zavialov P. O., Kostianoy A. G., Emelianov S. V., Ni A. A., Ishniyazov D., et al. (2003b) Hydrographic survey in the dying Aral Sea. *Geophysical Research Letters*, doi: 10.1029/2003GL017427, art.no.1659, **30**(13), 2-1– 2-4.

Zavialov P. O., Kostianoy A. G., Sapozhnikov Ph. V., Scheglov M. A., Khan V. M., Ni A. A., et al. (2003c) Present hydrophysical and hydrobiological state of the western Aral Sea. *Oceanology*, **43**(2), 316–319 [in Russian].

Zavialov P. O., Ginzburg A. I., Sapozhnikov Ph. V., Abdullaev U. R., Ambrosimov A. K., Andreev I. N., et al. (2004a) Interdisciplinary field survey of western Aral Sea in October, 2003. *Oceanology*, **44**(4), 667–670 [in Russian].

Zavialov P. O., Sapozhnikov Ph. V., and Ni A. A. (2004b) The gone sea. *National Geographic (Russian edition)*, **5**, 32–40 [in Russian].

Zavialov P. O., Amirov F. O., Dikarev S. N., and Rodin D. A. (2004c) Present Aral as an extreme example of metamorphization of a marine type water body under conditions of river discharge deficit. In: *Proceedings, VI Conference "Dynamics and Thermics of Rivers, Reservoirs, and Marine Coastal Zones"*. IVP RAS, Moscow, pp. 214–216 [in Russian].

Zhdanko S. M. (1940) Currents in the Aral Sea. *Meteorologiya i Gidrologiya*, **1–2**, 78–82 [in Russian].

Zubov N. N. (1947) *Dynamical Oceanology*. Gidrometeoizdat, Moscow, 360 pp [in Russian].

Index